Advances in Environmental, Economic and Social Assessment of Energy Systems

Advances in Environmental, Economic and Social Assessment of Energy Systems

Editor

Diego Iribarren

MDPI • Basel • Beijing • Wuhan • Barcelona • Belgrade • Manchester • Tokyo • Cluj • Tianjin

Editor
Diego Iribarren
IMDEA Energy
Spain

Editorial Office
MDPI
St. Alban-Anlage 66
4052 Basel, Switzerland

This is a reprint of articles from the Special Issue published online in the open access journal *Energies* (ISSN 1996-1073) (available at: https://www.mdpi.com/journal/energies/special_issues/Economic_Social_Energy).

For citation purposes, cite each article independently as indicated on the article page online and as indicated below:

LastName, A.A.; LastName, B.B.; LastName, C.C. Article Title. *Journal Name* **Year**, *Article Number*, Page Range.

ISBN 978-3-03943-450-3 (Hbk)
ISBN 978-3-03943-451-0 (PDF)

Contents

About the Editor

Diego Iribarren (Ph.D.) Dr. Diego Iribarren is a senior researcher in the Systems Analysis Unit of IMDEA Energy (Spain). He received his Ph.D. (2010) and B.Sc. (2005) degrees in Chemical and Environmental Engineering from the University of Santiago de Compostela (Spain). His current research activity focuses on the advanced analysis of energy systems under technical, economic, environmental, and social aspects from a life-cycle perspective. He has been involved in more than 40 research projects and contracts at both national and international levels, and has published more than 80 articles in peer-reviewed international journals, 10 book chapters, and more than 100 contributions to national and international conferences. He is currently one of the three chairs of the Spanish Network for Life Cycle Assessment (esLCA) and participates actively in international networks such as the European Energy Research Alliance and the International Energy Agency Hydrogen. He performs editorial tasks for several MDPI journals. He was awarded the I3 certificate from the Spanish State Research Agency in May 2018, and the Hydrogen Europe Young Scientist Award for the cross-cutting pillar in November 2018. He is one of the developers of the web-based software GreenH2armony® for the calculation of harmonised life-cycle indicators of hydrogen.

Preface to "Advances in Environmental, Economic and Social Assessment of Energy Systems"

Achieving a sustainable energy system is a global cornerstone of sustainable development. In this sense, technical advancements in the field of energy need to be complemented with a thorough assessment of their life-cycle performance in terms of sustainability. Thus, Life Cycle Assessment has become a reference methodology when it comes to evaluating the suitability of product systems, including energy systems. However, "advances in environmental, economic and social assessment of energy systems" are still needed to increase the robustness of energy systems analyses from a life-cycle sustainability perspective. This book brings together five contributions belonging to the field of energy systems analysis, with the common goal of contributing to the provision of sustainable energy solutions. These contributions encompass a wide range of aspects, from technical and environmental issues at the technology level (e.g., "adsorption capacity of organic compounds using activated carbons in zinc electrowinning", and "estimation of carbon dioxide emissions from a diesel engine powered by lignocellulose-derived fuel"), through economic parameters on cost-optimal energy solutions (regarding e.g., the "geographical potential of solar thermochemical jet fuel production" and "the economic and environmental effects of the cost-optimal energy renovation of a historic building district on the district heating system), to sustainability indicators of energy systems for multi-criteria decision analysis (concerning e.g., "wood-based bioenergy alternatives for residential heating"). Overall, this book contributes to enriching the literature on energy systems analysis in the path towards sustainable energy solutions.

Diego Iribarren
Editor

Article

Adsorption Capacity of Organic Compounds Using Activated Carbons in Zinc Electrowinning

Jung Eun Park [1], Eun Ju Kim [1], Mi-Jung Park [2] and Eun Sil Lee [1,*]

[1] Center for Plant Engineering, Institute for Advanced Engineering, Yongin-si 17180, Korea; jepark0123@gmail.com (J.E.P.); ejkim@iae.re.kr (E.J.K.)
[2] WESCO Electrode, Changwon-si, Gyeongsangnam-do 642370, Korea; pretty7008@naver.com
* Correspondence: les0302@iae.re.kr; Tel.: +82-31-330-7209

Received: 7 May 2019; Accepted: 3 June 2019; Published: 6 June 2019

Abstract: The influence of adsorbate (D2EPHA and kerosene) on the process of zinc electrowinning from sulfuric acid electrolytes was analyzed. The main objective was to critically compare three factors: (1) Three types of activated carbon (AC); (2) adsorption temperatures and contact time; and (3) zinc recovery efficiency. The results showed that organic components reduced the efficiency of zinc recovery. Moreover, wood-based ACs had a higher adsorption capacity than coal- and coconut-based ACs. To maintain a removal efficiency of 99% or more, wood-based ACs should constitute at least 60% of the adsorbate. The temperature of adsorption did not affect the removal efficiency. Additionally, the feeding rate of adsorbate in the solvent was inversely proportional to the removal efficiency. A feeding rate of the liquid pump of over 3 mL/min rapidly increased the delta pressure. For the same contact time, 99% of adsorbate removal occurred at 1 mL/min compared to approximately 97% at 0.5 mL/min. In the presence of 100 mg/L zinc, with increasing adsorbate from 0–5%, the recovery efficiency of zinc decreased from 100% to 0% and the energy consumption increased from 0.0017–0.003 kwh/kg zinc. Considering the energy consumption and zinc deposit mass, 0.1% of the adsorbate is recommended for zinc electrowinning.

Keywords: zinc (Zn); electrowinning (EW); activated Carbons (ACs); adsorbate; liquid phase space velocity (LHSV); temperature

1. Introduction

Electrowinning (EW) of rare metals, such as zinc (Zn), copper (Cu), and manganese (Mg), has been widely used due to its low energy consumption and high output [1–5]. The process of Zn electrowinning is conducted in several stages, including solvent extraction (organic mixing and separation), Zn removal, and electrolytic winning [6,7]. Solvent extraction is a particularly important unit operation in the purification and concentration of these materials. Several researchers have studied solvent extraction from an aqueous leaching solution using organic extractants [6–14]. Generally, the extraction of Zn and selected base metal and alkali cations to produce organophosphorus-based extractants; i.e., phosphoric-acid extractants (de(2-ethylhexyl)phosphoric acid(D2EHPA), phosphonic-acid extractant (Ionquest 801), and phosphinic-acid extractants (CYANEX 272), depends on the pH, O/A phase ratio, and concentration. At first, Devi et al. investigated the extraction of Zn from sulphate solution using sodium salts in kerosene and compared the effect of pH, extraction concentration and various sodium salts [11]. Among the organophosphorus-based extractants, D2EHPA has shown the best extraction efficiency for increasing Zn impurities [11,14,15]. The effects of several parameters on Zn extraction from phosphoric acid solution were found to have the following order of importance: D2EHPA concentration > equilibrium pH > O/A phase ratio [16].

The presence of organophosphorus-based extractants decreases metal impurities, including the concentration of Zn [17]. Ivanov's research group focused on Zn impurities through the addition of inhibitors during electrowinning [18,19]. They reported that the addition of inhibitors to the electrolytes caused Zn re-dissolution. For this reason, it is necessary to explore effective methods for the efficient removal of organic components in sulfuric acid solvent components to improve the Zn electrowinning process. The adsorption of organic components from sulfuric acid by carbon has been studied for several decades and is becoming more widespread, due to large surfaces and strong adsorption [19–22]. Hydrophobic carbons are more effective adsorbents for trichloroethene (TCE) and methyl tertiary-butyl ether (MTBE) than hydrophilic carbons because enhanced water adsorption on the latter interferes with the adsorption of micropollutants from solutions containing natural organic matter [22]. Among all types of carbon, activated carbon (AC) is generally considered to have a strong adsorption affinity for organic chemicals, due to their highly hydrophobic surfaces. With respect to pore structure, optimal AC should exhibit a large volume of micropores approximately 1.5 times the kinetic diameter of the target adsorbate [22].

Therefore, in order to determine an efficient removal method of organic components in sulfuric acid solvent components, it is necessary to determine the uppermost limit of organic compounds that will not affect the Zn electrowinning efficiency. Therefore, this study has three principle objectives: (1) To compare the performance of three types of AC as an adsorbent; (2) to investigate the effects of adsorption temperature and contact time; and (3) to determine the efficiency of Zn recovery with organic components.

2. Materials and Methods

2.1. Adsorbate

The D2EHPA extractants were provided by Mining Chem Co., Ltd., Seoul, Korea, and dissolved in treated kerosene (ESCAID 110). The kerosene used as the diluent in this study was a commercial ESCAID 110 product from Minning CAM. We prepared 2M D2EPHA using kerosene (the ratio of D2EPHA and kerosene was 7:3). The extractants were added to the electrolytes in various proportions from 0.00% to 5.00% in sulfuric acid as an electrolyte.

2.2. Activated Carbons (ACs) as Adsorbents

The commercially ACs were obtained adsorbents, as shown in Table 1. To compare the ACs, wood-based AC (Wood-AC, JCG-10, Ja Yeon Science Ind. Co., Chulwon, Korea), coal-based AC (Coal-AC, NCL, Neven Ind. Co., Pohang, Korea), and coconut-based AC (Coconut-AC, Ya-1, Yeon Science Ind. Co., Chulwon, Korea) were tested. Before adsorption, ACs was dried at 110 °C and stored in a desiccator.

The most common AC characteristics reported in previous adsorption literature are its specific surface area, total pore volume, and micropore volume. Surface area and total pore volume were determined from N_2 isotherm data collected at 196 °C, which were measured using an adsorption analyzer (ASAP-2010, Micromeritics Inc., Norcross, GA, USA). The Brunauer-Emmet-Teller (BET) theory was used to determine the specific surface area and the total pore volume was calculated from the amount of N_2 gas adsorbed at a relative pressure of 0.95. The Horváth–Kawazoe (HK) method was applied to calculate the micropore volume. Prior to analysis, AC samples were outgassed overnight at 110 °C.

For the proximate analysis, dried samples were placed in a furnace (Daeheung Science, DF-4S, Incheon, Korea) and heated at 950 °C for 7 min. The weight of the samples was measured to determine the volatile matter content. The samples were placed in the furnace again and heated at 750 °C for 10 h to measure the amount of ash. The ash, volatile matter, and fixed C contents within the ACs were reported as a weight percentage. Elemental contents of carbon (C), hydrogen (H), oxygen (O), nitrogen (N), and sulfur (S) were determined with an elemental analyzer (FLASH 2000, Thermo Fisher Scientific, Waltham, MA, USA).

Table 1. The list of ACs as adsorbents and their textural characteristics.

Adsorbents		Wood-AC	Coal-AC	Coconut-AC
Raw Material		Wood	Coal	Coconut
Surface Area (m^2/g)		1398	1030	1067
Total Pore volume (m^3/g)		1.19	0.52	0.45
Micropore size (A)		6.56	6.16	5.38
Proximate analysis (wt.%)	Moisture	0.18	0.18	0.72
	Volatile	2.24	2.24	2.52
	Fixed Carbon	72.70	88.62	94.28
	Ash	20.90	8.96	2.48
Ultimate analysis (wt.%)	Carbon	68.8	88.2	94.2
	Hydrogen	1.0	0.4	0.4
	Oxygen	15.7	0.5	2.3
	Nitrogen	0.3	0.3	2.0
	Sulphur	0.1	0.0	0.0

2.3. Adsorption Test

We compared the continuous stirred tank reactor (CSTR) and the packed bed reactor (1PBR and 2PBR) adsorption methods, as shown in Table 2. For the CSTR method, 100 mL solvent of 3 M H_2SO_4 was added to the AC adsorbate and stirred for 1 h at 300 rpm. After filtration, we analyzed the organic components. For the PBR method, 0.5–1.5 g of AC was loaded into a $\frac{1}{4}$-inch glass reactor and the solvent was fed using a liquid pump (Model 781100, KD Scientific, Holliston, MA, USA). The flow rate was set to 0.5, 1.0, 3.0, and 5.0 mL/min.

Table 2. The scheme of reactors.

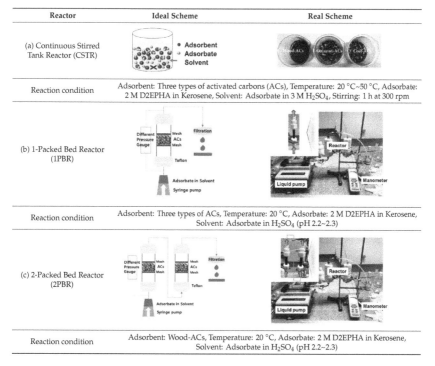

Reactor	Ideal Scheme	Real Scheme
(a) Continuous Stirred Tank Reactor (CSTR)	• Adsorbent / ◦ Adsorbate / Solvent	
Reaction condition	Adsorbent: Three types of activated carbons (ACs), Temperature: 20 °C~50 °C, Adsorbate: 2 M D2EPHA in Kerosene, Solvent: Adsorbate in 3 M H_2SO_4, Stirring: 1 h at 300 rpm	
(b) 1-Packed Bed Reactor (1PBR)		
Reaction condition	Adsorbent: Three types of ACs, Temperature: 20 °C, Adsorbate: 2 M D2EPHA in Kerosene, Solvent: Adsorbate in H_2SO_4 (pH 2.2~2.3)	
(c) 2-Packed Bed Reactor (2PBR)		
Reaction condition	Adsorbent: Wood-ACs, Temperature: 20 °C, Adsorbate: 2 M D2EPHA in Kerosene, Solvent: Adsorbate in H_2SO_4 (pH 2.2~2.3)	

3

2.4. Adsorbate Analysis

2.4.1. n-Hexane Extraction

The adsorbate was extracted with 100 mL of solvent and shaken twice over 5 min. After separation, the extracted adsorbate in n-Hexane was boiled at 80 °C. The adsorbate contents yielded was expressed in terms of the mass percentage of the samples. The extracted adsorbate yield can be estimated using the following Equation (1),

$$\text{Extracted Adsorbate Yield (wt.\%)} = \frac{\text{Mass of extracted adsorbate (g)}}{\text{Mass of adsorbate (g)}} \times 100 \qquad (1)$$

2.4.2. Total Organic Carbon (TOC)

Total organic carbon (TOC) was evaluated using TOC cell kits (Merck, Darmstadt, Germany) which is analogous to the APHA 5310 C (2014) method. Ten milliliters of the sample was prepared for the TOC analysis. The solution was pretreated with a reagent containing sulfuric acid according to the specifications of the Merck kits, titrated to pH 2.2–2.3, and stirred at a low speed for 10 min. In order to remove a small amount of AC, filtration was carried out with a 0.2-μm syringe filter. Three milliliters of sample was added to the cell kit. Immediately after treatment with a reagent containing peroxydisulfate, the cell was tightly closed with an aluminum crew cap then stood upside down and heated at 120 °C using a heating block for 2 h. After cooling for 1 h, it was analyzed by Spectroquant® (Merck, Kenilworth, NJ, USA).

2.5. Characterization of Electrodes

Zn electrowinning experiments were conducted in 3 M H_2SO_4 solution with adsorbate using ZIVE (MP2PC) from WonA Tech. (Seoul, Korea). The cell consisted of an aluminum anode (2 × 2.5 cm²) and Pd-Ir-Sn-Ta/TiO_2 cathode (2 × 2.5 cm²). The anode–cathode distance was 3 cm, as shown in Figure 1a. A commercial electrode, 8.6 g Pd-Ir-Sn-Ta/TiO_2, was purchased from West Co. (Changwon-si, Korea) and the substrate was an aluminum plate. The metal composition was obtained using energy-dispersive X-ray analysis (EDAX, Inspect F50, Thermo Fisher Scientific, Waltham, MA, USA) and the electrode compositions of Pd, Ir, Sn, and Ta were 8.6, 39.9, 18.5, and 33.0% [23]. The thickness of coating materials was 7.6~8.8 um and is well dispersed, as shown in Figure 1b,c.

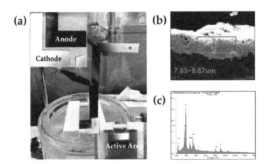

Figure 1. Electrowinning test cell setup and electrode properties. (**a**) The electrowinning reactor consisted of a 1 l double jacket (insert Figure showed that 8.6g Pd-Ir-Sn-Ta/TiO_2 as anode and aluminum plate as cathode). (**b**) The micrograph and (**c**) energy-dispersive X-ray analysis (EDAX) pattern of the 8.6 g Pd-Ir-Sn-Ta/TiO_2 electrode.

For the Zn electrowinning, the electrolyte (100 g/L Zn and 3 M H_2SO_4) was prepared using sulfuric acid and zinc oxide. After electrolyte preparation, the adsorbate was added in proportions from 0.0–5.0%. The electrowinning reactor consisted of a 1 l double jacket, so a reaction temperature of 40 °C was maintained using bath circulator throughout. The electrowinning measurements were carried out at a current density of 500 A/dm^2 for 4 h.

3. Results and Discussion

3.1. Characterization of ACs

A list of ACs and their surface area, total pore volume, and micropore size values are presented in Table 1. Wood-AC had high volatile and ash matter content and low moisture content. According to the ultimate analysis, the carbon content of Coconut-ACs was higher than that of Coal-ACs and Wood-ACs. Moreover, the oxygen content decreased in the following order: Wood-ACs > Coconut-AC > Coal-AC (Table 1). All three commercial ACs contained approximately zero sulfur, decreasing the potential for acid species formation.

Figure 2 shows the adsorption/desorption isotherms of nitrogen obtained for all three ACs analyzed in this study. N$_2$ adsorption and desorption isotherms of Wood-ACs followed the trend of type IV isotherms (Figure 2a), whereas the isotherms of Coal-ACs and Coconut-ACs belonged to type I isotherms according to the IUPAC classification.

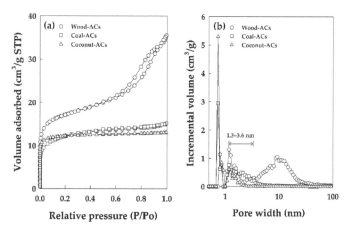

Figure 2. Physical properties of ACs (○: Wood-ACs, □: Coal-ACs, and △: Coconut-ACs). (**a**) Nitrogen ad-/de-sorption isotherms and (**b**) pore size distribution for ACs.

Furthermore, Figure 2b illustrates the pore size distribution of ACs. The pore widths of Wood-ACs were higher than those of Coal-ACs and Coconut-ACs (Table 1). The surface area of Wood-ACs was higher than that of Coal-ACs and Coconut-ACs. Total pore volume and micropore size (<2 nm) decreased in the following order: Wood-ACs > Coal-ACs ≥ Coconut-ACs. In addition, the percentages of mesopores (2–50 nm) for Wood-ACs, Coal-ACs, and Coconut-ACs were 85, 13, and 4%, respectively.

3.2. Adsorbate Removal Efficiency

To evaluate the effect of AC types (Wood-ACs, Coal-ACs, and Coconut-ACs) on adsorption capacities in the presence of adsorbate, we conducted an adsorbate test. As shown in Table 3, the adsorbate adsorption capacities of Wood-ACs, Coal-ACs, and Coconut-ACs were 99%, 98%, and 93%, respectively, with Wood-ACs exhibiting the largest adsorption capacity for adsorbate, including the highest BET surface area and total pore volume. Quinlivan et al. compared the effects of physical and chemical activated carbon on adsorption capacities; they exhibited a large volume of micropores with widths approximately 1.3 to 1.8 times larger than the kinetic diameter of the target adsorbent [22]. Considering the molecular weight of D2EPHA and kerosene as the adsorbate, the pore diameter is predicted to be at least 10 A to 20 A [24,25]. AC, which is effective at adsorbing organic matter, has a large number of pores with diameters of at least 13 A to 36 A. Therefore, the large surface area and mesopore volume of Wood-ACs indicates their high removal efficiency.

Table 3. The removal efficiency for investigated ACs.

Adsorbents		Wood-ACs	Coal-ACs	Coconut-ACs
Removal efficiency (%)	N-Hexane [1]	99.76	98.83	92.81
	Ave. TOC [2]	98.29	97.45	95.10

[1] CSTR conditions (Adsorbents—5.0 g, Adsorbate contents—5.0 g in 100 mL of 3 M H_2SO_4 solvent, stirring speed—300 rpm); [2] PBR conditions (flow rate—1.0 mL/min, Adsorbents—1.0 g, Adsorbate—1.0 g in 100 mL of pH 2.2–2.3 solvent (H_2SO_4 with D.I water).

Apart from textual characteristics, the behavior of ACs is often strongly influenced by oxygen, which affects the surface hydrophobicity [19,26–29]. The surface hydrophobicity of carbon materials plays an important role in interface and colloid science. An increase in the oxygen content of carbon leads to a decrease in its hydrophobicity. Thus, the adsorption of adsorbate on ACs (Coal-ACs and Coconut-ACs) decreased when the oxygen content of the carbonaceous adsorbent increased.

The removal efficiency of adsorbate components was analyzed in detail using the PBR system. The removal efficiency trends for the PBR system were the same as those of the CSTR system; Wood-ACs > Coal-ACs > Coconut-ACs. When adsorbing 600 mg of adsorbate, Wood-ACs exhibited residual adsorbate values of 5.7 mg, whereas Coal-ACs and Coconut-ACs showed residual adsorbate values of 8.6 mg and 16.5 mg, respectively.

Among the ACs, Wood-ACs showed the highest removal efficiency of adsorbate (Figure 3). Therefore, we compared the removal efficiency of adsorbate for the different weights of Wood-AC (0.1 g to 5.0 g). As shown in Figure 4, the removal efficiency of adsorbate depended on the ratio of adsorbents; the removal efficiency increased with decreasing adsorbate components. It should be noted that the removal efficiency results show a very good correlation. Using the n-Hexane method, ACs should utilize at least 60% of the sorbent to remove 99% of the adsorbate before the AC breaking point occurs.

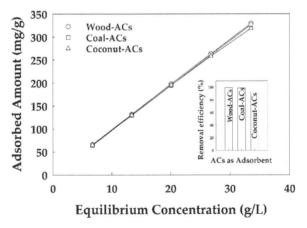

Figure 3. The adsorbed amount as equilibrium concentration depends on a different type of ACs viz. Wood-ACs, Coal-ACs, and Coconut-ACs as PBR conditions, as shown in Table 3. (insert Figure means that the removal efficiency of adsorbate as ACs type and detected adsorbate after adsorption analyzed by n-Hexane method. CSTR conditions, as shown in Table 3).

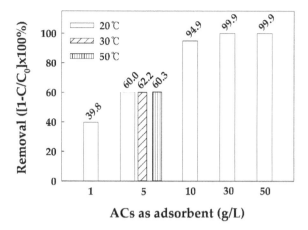

Figure 4. The removal efficiency of adsorbate as adsorbent contents and adsorption temperature of 20–50 °C. Experimental conditions—CSTR, Adsorbents—5.0 g, Adsorbate contents—5.0 g in 100 mL of 3 M H_2SO_4 solvent, stirring speed—300 rpm).

Generally, the adsorption temperature is thought to be important in the gas phase. However, an adequate adsorption temperature has yet to be proven in the liquid phase for ACs. To assess the effect of adsorption temperature on the adsorbent, Figure 4 compares the removal efficiency of Wood-AC adsorbate at three temperatures (from 20–50 °C). Absorption of the adsorbate within this temperature range is included within the error range. In previous research, they compared the adsorption isotherms of diclofenac onto ACs at different operation temperature (30–60 °C). The mass of adsorbed diclofenac was less dependent on the temperature at high concentration of diclofenac [30].

We also investigated the influence of flow rate, removal efficiency, and delta pressure (P) at a liquid hourly space velocity (LHSV) of 0.005–0.026 h^{-1}. As shown in Figure 5, the adsorbate adsorption capacities were largest at lower space velocity; i.e., 0.005 h^{-1} > 0.016 h^{-1} > 0.026 h^{-1}. Moreover, the space velocity increased with increasing delta P. To achieve 99.5% adsorbate removal from Wood-ACs, a space velocity of approximately 0.016 h^{-1} was required. However, delta P should be maintained under 1 atm at the adsorption condition; otherwise, the elimination efficiency of the adsorbate is reduced, and the adsorbate adsorbed onto the surface of ACs is thought to be detoxified.

To design the factors of the adsorption system, we experimentally compared the efficiency of adsorbate removal under the same space velocity conditions. The first reaction condition supplied the adsorbate at 0.5 mL/min and employed 1PBR. The second reaction condition supplied the adsorbate at 1.0 mL/min and employed 2PBR.

According to the results, up to 250 g/L of equilibrium concentration was adsorbed under both conditions. However, differences appeared in the removal efficiency of the adsorbate at greater than 250 g/L of equilibrium concentration (Figure 6). Therefore, in order to effectively eliminate the adsorbate, it is more effective to increase the number of PBR than to regulate the supply speed of the adsorbate.

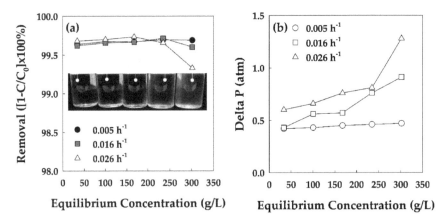

Figure 5. The adsorbate removal efficiency and delta pressure at a liquid hourly space velocity of 0.005–0.026 h^{-1}. (**a**) The removal efficiency as space velocity of 0.005–0.026 h^{-1} (insert Figure showed the solvents after adsorption test). (**b**) Delta pressure as space velocity of 0.005–0.026 h^{-1}. Experimental conditions—1.0 g of ACs and flow rate of solvent was set to 1.0, 3.0, and 5.0 mL/min.

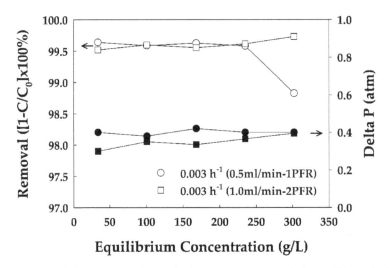

Figure 6. Removal efficiency depends on adsorbents at same space velocity conditions. (○: The flow rate of solvent is 0.5 mL/min and the weight of adsorbent is 1.0 g, □: The flow rate of solvent is 1.0 mL/min and the weight of adsorbent is 2.0 g).

3.3. Zn Electrowinning Test

This adsorbate adsorption test revealed the factors necessary for the efficient recovery of high purity Zn during the pre-treatment process. Therefore, if the adsorbate was not efficiently removed during the pre-treatment process, the effects of the Zn recovery process were considered. Thus, a Zn recovery experiment was conducted using an adsorbate concentration range from 0.0–5.0%.

As shown in Figure 7, the Zn recovery efficiency tended to decrease as the adsorbate content increased. An electrolyte containing more than 2% adsorbate led to a rapid reduction in Zn recovery efficiency. In the case of 0% adsorbate, the voltage was maintained at 2.7 V for 4 h. For 0.1–1.0% adsorbate in the electrolyte, the initial voltage fluctuated for 0.5–1 h. The average voltage for 4 h increased as increasing adsorbate concentration. As increased the adsorbate concentration of 0.00, 0.10, 0.30, 0.50, 1.00%, the average voltage increased 2.7972 V, 2.9874 V, 3.0109 V, 3.0466 V, and 3.0728 V, respectively. In other words, as increased the adsorbate concentration from 0.00 to 1.00% was increasing consumption energy for Zn recovery from 0.0017 kWh to 0.0031 kWh. According to the previous research, the current efficiency and impurity were found to decrease with increasing additive concentration. Also, the voltage (2.80–2.95) and energy consumption (2675–2799 kWh/t) increase with increasing concentration of additives (0–15ppm) [30]. In the presence of organic from 0 to 100 mg/L, the current efficiency decreased from 93 to 61% [31]. As discussed later in this paper, this organic caused significant reductions in the current efficiency of the electrowinning process, thereby increasing energy consumption. At more than 2% adsorbate, the voltage fluctuated rapidly and hard to verify a stable voltage. Therefore, as the adsorbate increased, the consumption energy increased, and Zn recovery efficiency decreased.

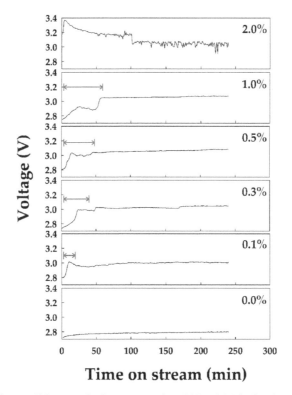

Figure 7. Voltage as adsorbate contents from 0.0% to 2.0% for four hours.

As shown in Figure 8a, Zn recovery efficiency in the presence of adsorbate concentration from 0.00 to 0.50% was recovered at least 99.9%. In the presence of more than 0.50% of adsorbate, Zn recovery efficiency tends to rapidly decrease. The surface of an electrode recovered from an electrolyte with different adsorbate contents, as shown in Figure 8b. In the absence of adsorbate, the electrode surface exhibited uniform deposition. However, for 0.1–2% adsorbate, the electrode surface appeared to have grown very unevenly. The addition of the organics to the electrolyte changed the features of the metal deposit [31]. From the microscopy analysis, the addition of organics to the electrolyte leads to the formation of pore on the deposition surface. Therefore, the presence of adsorbate reduces Zn recovery efficiency and leads to non-uniform deposition. Thus, adsorbate removal must be less than 0.1%.

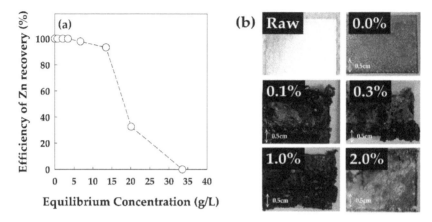

Figure 8. Energy efficiency of Zn recovery and images after Zn electrowinning. (**a**) Zn recovery efficiency as adsorbate concentration and (**b**) the electrode surface images after Zn recovery.

4. Conclusions

The objective of this research was to determine the effect of adsorbate on the electrolyte in the Zn electrowinning process. Experiments were conducted to determine the effects of three different parameters: (1) Adsorbent type, (2) adsorbate content, and (3) Zn recovery.

Among the three different types of activated carbon (AC), wood-based AC (Wood-AC) showed the highest adsorbate adsorption capacity and is thought to be more suitable for adsorbate adsorption, due to its surface area and pore size. If Wood-ACs with a surface area of 1000 m^2/g account for at least 60% of the adsorbate, more than 99% is eliminated. Moreover, the adsorption capacity did not differ significantly within a temperature range of 20–50 °C.

The Zn recovery efficiency tended to show an inversely proportional relationship to the amount of adsorbate. In other words, the Zn recovery efficiency decreased from 100% to 0% when the adsorbate content increased from 0% to 5%. In addition, as the adsorbate content increased, the voltage increased from 2.7 V to 3.1 V; thus, the consumption energy increased from 0.0017 to 0.0031 kWh. Therefore, considering the energy consumption and Zn deposit mass, 0.1% of adsorbate is recommended for Zn electrowinning.

Author Contributions: Conceptualization, J.E.P. and E.S.L.; methodology, J.E.P. and E.J.K.; formal analysis, J.E.P. and E.J.K.; writing—original draft preparation, J.E.P.; writing—review and editing, J.E.P. and E.S.L.; project administration, E.S.L.; funding acquisition, M.-J.P.

Funding: This study was supported by the Energy Development Technology Program of the Korea Institute of Energy Technology Evaluation and Planning (KETEP) granted financial resources from the Ministry of Trade, Industry and Energy, Korea (20172010105220) and also the National Research Foundation of Korea (NRF) and the Center for Women In Science, Engineering and Technology (WISET) Grant funded by the Ministry of Science and ICT under the Program for Returners into R&D.

Conflicts of Interest: The authors declare no conflict of interest.

References

1. Gladysz, O.; Los, P.; Krzyzak, E. Influence of concentrations of copper, leveling agents and temperature on the diffusion coefficient of cupric ions in industrial electro-refining electrolytes. *J. Appl. Electrochem.* **2007**, *37*, 1093–1097. [CrossRef]
2. Moats, M.; Free, M. A bright future for copper electrowinning. *JOM* **2007**, *59*, 34–36. [CrossRef]
3. Alfantazi, A.; Valic, D. A study of copper electrowinning parameters using a statistically designed methodology. *J. Appl. Electrochem.* **2003**, *33*, 217–225. [CrossRef]

4. Xue, J.; Wu, Q.; Wang, Z. Function of additives in electrolytic preparation of copper powder. *Hydrometallurgy* **2006**, *82*, 154–156. [CrossRef]

5. Muresan, L.; Nicoara, A.; Varvara, S. Influence of Zn^{2+} ions on copper electrowinning from sulfate electrolytes. *J. Appl. Electrochem.* **1999**, *29*, 719–727. [CrossRef]

6. Cole, P.M.; Sole, K.C. Zinc solvent extraction in the process industries. *Miner. Process. Extr. Metall. Rev.* **2003**, *24*, 91–137. [CrossRef]

7. Zhu, Z.; Cheng, C.Y. A study on zinc recovery from leach solutions using Ionquest 801 and its mixture with D2EHPA. *Miner. Eng.* **2012**, *39*, 117–123. [CrossRef]

8. Jha, M.K.; Gupta, D.; Choubey, P.K.; Kumar, V.; Jeong, J.; Lee, J. Solvent extraction of copper, zinc, cadmium and nickel from sulfate solution in mixer settler unit (MSU). *Sep. Purif. Technol.* **2014**, *122*, 119–127. [CrossRef]

9. Daryabor, M.; Ahmadi, A.; Zilouei, H. Solvent extraction of cadmium and zinc from sulphate solutions: Comparison of mechanical agitation and ultrasonic irradiation. *Ultrason. Sonochem.* **2017**, *34*, 931–937. [CrossRef]

10. Verbeken, K.; Verhaege, M.; Wettinck, E. Separation of iron from zinc sulfate electrolyte by combined liquid-liquid extraction and electroreductive stripping. In *Lead-Zinc 2000*; Dutrizac, J.E., Gonzalez, J.A., Henke, D.M., James, S.E., Siegmund, A.H.-J., Eds.; The Minerals, Metals & Materials Society: Warrendale, PA, USA, 2013; pp. 779–788. [CrossRef]

11. Devi, N.B.; Nathsarma, K.C.; Chakravortty, V. Solvent extraction of zinc(II) using sodium salts of D2EHPA, PC88A and Cyanex 272 in kerosene. In Proceedings of the Mineral Processing: Recent Advances and Future Trends, Indian Institute of Technology, Kanpur, India, 11–15 December 1995; Mehrotra, S.P., Rajiv, S., Eds.; Volume 11–15, pp. 537–547.

12. Jha, M.K.; Kumar, V.; Jeong, J.; Lee, J. Review on solvent extraction of cadmium from various solutions. *Hydrometallurgy* **2012**, *111–112*, 1–9. [CrossRef]

13. Deep, A.; de Carvalho, J.M.R. Review on the Recent Developments in the Solvent Extraction of Zinc. *Solvent Extr. Ion Exch.* **2008**, *26*, 375–404. [CrossRef]

14. Nathsarma, K.C.; Devi, N.B. Separation of Zn(II) and Mn(II) from sulphate solutions using sodium salts of D2EHPA, PC88A and Cyanex 272. *Hydrometallurgy* **2006**, *84*, 149–154. [CrossRef]

15. Asadi, T.; Azizi, A.; Lee, J.; Jahani, M. Solvent extraction of zinc from sulphate leaching solution of a sulphide-oxide sample using D2EHPA and Cyanex 272. *J. Dispers. Sci. Technol.* **2017**, *39*, 1328–1334. [CrossRef]

16. Mellah, A.; Benachour, D. The solvent extraction of zinc and cadmium from phosphoric acid solution by di-2-ethyl hexyl phosphoric acid in kerosene diluent. *Chem. Eng. Process.* **2006**, *45*, 684–690. [CrossRef]

17. Dhak, D.; Asselin, E.; Carlo, S.D.; Alfantazi, A. An investigation on the effects of organic additives on zinc electrowinning from industrial electrolyte. *ECS Trans.* **2010**, *28*, 267–280. [CrossRef]

18. Ivanov, I. Increased current efficiency of zinc electrowinning in the presence of metal impurities by addition of organic inhibitors. *Hydrometallurgy* **2004**, *72*, 73–78. [CrossRef]

19. Moreno-Castilla, C. Adsorption of organic molecules from aqueous solutions on carbon materials. *Carbon* **2004**, *42*, 83–94. [CrossRef]

20. Apul, O.G.; Karanfil, T. Adsorption of synthetic organic contaminants by carbon nanotubes: A critical review. *Water Res.* **2014**, *68*, 34–55. [CrossRef]

21. Zhang, S.; Shao, T.; Bekaroglu, S.S.K.; Karanfil, T. Adsorption of synthetic organic chemicals by carbon nanotubes: Effects of background solution chemistry. *Water Res.* **2010**, *44*, 2067–2074. [CrossRef]

22. Quinlivan, P.A.; Li, L.; Knappe, D.R.U. Effects of activated carbon characteristics on the simultaneous adsorption of aqueous organic micropollutants and natural organic matter. *Water Res.* **2005**, *39*, 1663–1673. [CrossRef]

23. Park, J.E.; Yang, S.K.; Kim, J.H.; Park, M.-J.; Lee, E.S. Electrocatalytic activity of $Pd/Ir/Sn/Ta/TiO_2$ composite electrodes. *Energies* **2018**, *11*, 3356. [CrossRef]

24. Dora, S.K. Real time recrystallization study of 1, 2dodecanediol on highly oriented pyrolytic graphite (HOPG) by tapping mode atomic force microscopy. *WJNSE* **2017**, *7*, 1–15. [CrossRef]

25. Lu, J.R.; Thomas, R.K.; Binks, B.P.; Fletcher, P.D.I.; Penfold, J. Structure and composition of dodecane layers spread on aqueous solutions of dodecyland hexadecyltrimethylammonium bromides studied by neutron reflection. *J. Phys. Chem.* **1995**, *99*, 4113–4123. [CrossRef]

26. Pendleton, P.; Wu, S.H.; Badalyan, A. Activated carbon oxygen content influence on water and surfactant adsorption. *J. Colloid Interface Sci.* **2002**, *246*, 235–240. [CrossRef] [PubMed]

27. Kim, J.-H.; Wu, S.H.; Pendleton, P. Effect of surface properties of activated carbons on surfactant adsorption kinetics. *Korean J. Chem. Eng.* **2005**, *22*, 705–711. [CrossRef]

28. Fendleton, P.; Wong, S.H.; Schumann, R.; Levay, G.; Denoyel, R.; Rouquerol, J. Properties of activated carbon controlling 2-methylisoborneol adsorption. *Carbon* **1997**, *35*, 1141–1149. [CrossRef]

29. Nam, S.-W.; Choi, D.-J.; Kim, S.-K.; Herc, N.; Zoh, K.-D. Adsorption characteristics of selected hydrophilic and hydrophobic micropollutants in water using activated carbon. *J. Hazard. Mater.* **2014**, *270*, 144–152. [CrossRef] [PubMed]

30. Tomul, F.; Arslan, Y.; Basoglu, F.T.; Babuccuoglu, Y.; Tran, H.N. Efficient removal of anti-inflammatory from solution by Fe-containing activated carbon: Adsorption kinetics, isotherms, and thermodynamics. *J. Environ. Manag.* **2019**, *238*, 296–306. [CrossRef]

31. Majuste, D.; Bubani, F.C.; Bolmaro, R.E.; Martins, E.L.C.; Cetlin, P.R.; Ciminelli, V.S.T. Effect of organic impurities on the morphology and crystallographic texture of zinc electrodeposits. *Hydrometallurgy* **2017**, *169*, 330–338. [CrossRef]

Article

Multi-Criteria and Life Cycle Assessment of Wood-Based Bioenergy Alternatives for Residential Heating: A Sustainability Analysis

Mario Martín-Gamboa [1,*], Luis C. Dias [2,3], Paula Quinteiro [1], Fausto Freire [4], Luís Arroja [1] and Ana Cláudia Dias [1]

[1]   Centre for Environmental and Marine Studies (CESAM), Department of Environment and Planning, University of Aveiro, Campus Universitário de Santiago, 3810-193 Aveiro, Portugal; p.sofia@ua.pt (P.Q.); arroja@ua.pt (L.A.); acdias@ua.pt (A.C.D.)

[2]   CeBER and Faculty of Economics, University of Coimbra, Av. Dias da Silva 165, 3004-512 Coimbra, Portugal; lmcdias@fe.uc.pt

[3]   INESCC—Institute for Systems Engineering and Computers at Coimbra, 3030-290 Coimbra, Portugal

[4]   ADAI-LAETA, Department of Mechanical Engineering, University of Coimbra, Polo II Campus, R. Luís Reis Santos, 3030-788 Coimbra, Portugal; fausto.freire@dem.uc.pt

*   Correspondence: m.martin@ua.pt; Tel.: +351-234370349

Received: 18 September 2019; Accepted: 18 November 2019; Published: 19 November 2019

Abstract: Moving towards a global bioeconomy can mitigate climate change and the depletion of fossil fuels. Within this context, this work applies a set of multi-criteria decision analysis (MCDA) tools to prioritise the selection of five alternative bioenergy systems for residential heating based on the combination of three commercial technologies (pellet, wood stove and traditional fireplace) and two different feedstocks (eucalypt and maritime pine species). Several combinations of MCDA methods and weighting approaches were compared to assess how much results can differ. Eight indicators were used for a sustainability assessment of the alternatives while four MCDA methods were applied for the prioritisation: Weighted Sum Method (WSM), Technique for Order of Preference by Similarity to Ideal Solution (TOPSIS), Elimination and Choice Expressing Reality (ELECTRE), and Preference Ranking Organization Method for Enrichment Evaluation (PROMETHEE). Regarding the sustainability performance indicators, the highest environmental impacts were calculated for the fireplace alternatives, and there was not a best environmental option. Also, no clear trend was found for the economic and social dimensions. The application of MCDA tools shows that wood stove alternatives have the best sustainability performance, in particular wood stove with combustion of maritime pine logs (highest scores in the ranking). Regarding the worst alternative, fireplaces with combustion of eucalypt logs ranked last in all MCDA rankings. Finally, a sensitivity analysis for the weighting of the performance indicators confirmed wood stoves with combustion of maritime pine logs as the leading alternative and the key role of the analysts within this type of MCDA studies.

Keywords: bioeconomy; life cycle assessment; multi-criteria decision analysis; sustainability; thermal energy; wood

---

1. Introduction

The effects of climate change and the continuous depletion of fossil fuels require a change in the global economy that should be based on the use of renewable resources. The implementation of processes based on the production, supply and processing of biomass can boost that change, reducing environmental impacts and ensuring the conservation of finite resources. Thus, the concept of bioeconomy emerges as a suitable opportunity to link economic growth with sustainable development [1,2]. The European Commission defines the bioeconomy concept as "the production of renewable biological resources and the conversion of these resources and waste streams into value added products, such as food, feed, bio-based products, and bioenergy" [3].

The bioeconomy created a turnover of 2.3 trillion € in the European Union (EU) in 2015, with the bio-based electricity and forestry sectors having the greatest growth rates [4]. Thus, the correct exploitation of forest resources can contribute to provide a sustainable source of energy, promoting the development of rural areas and jobs creation [5]. In particular, bioenergy options based on wood feedstocks are presented as a convenient solution for residential heating, especially in countries like Portugal where forest resources are still abundant [6–8]. Several wood-based technologies are currently available in the market for residential heating, from traditional fireplaces to wood/pellet stoves [9–11]. An appropriate selection of these technologies in terms of sustainability will be crucial for paving the way to a future wide deployment of the European bioeconomy.

When taking into consideration the sustainability of energy systems, the search for logical and optimal solutions is a complex process since it involves: (i) many sources of uncertainty at different scales (e.g., at technology, company or policy level), (ii) probably long time frames according to the lifespan of energy systems, (iii) capital intensive investments, and (iv) a large number of stakeholders with different views and preferences [12]. In this sense, multiple criteria are needed to reflect the complexity of the sustainability assessment for decision-makers and to create an adequate basis for a quality and comprehensive decision. This means taking into account not only environmental criteria but also economic and social indicators (covering the main dimensions of the sustainability concept). In this respect, approaches with a life-cycle perspective are considered as effective sources for supplying performance indicators in one dimension or in all dimensions of sustainability [13,14].

In addition to indicators, methods are needed to reduce the complexity of integrating and interpreting multiple criteria and preferences of stakeholders. Multi-criteria decision analysis (MCDA) arises as an operational evaluation and decision support approach that is suitable for addressing complex problems (e.g., energy and environmental issues) featuring high uncertainty, conflicting objectives, different forms of data and information, multiple interests and perspectives, and the accounting for complex and evolving biophysical and socio-economic systems [15,16]. According to Ibáñez-Forés et al. [17], MCDA methods can be classified into two broad groups: Multi-Attribute Decision Analysis (MADA) and Multi-Objective Decision Analysis (MODA). The former is used when the decision-maker has to choose between a finite number of options, while the latter implies setting the value of decision variables (implicitly defining an infinite number of alternatives). Within the MADA, which fits better with the interest of the present study, the most popular are the following: Multi-Attribute Utility Theories (MAUT), the Outranking methods, Analytical Hierarchy Process (AHP) method, and other multiple attribute decision-making (OMADM) methods (such as distance-to-target approaches) [12,18,19].

MCDA tools have been widely applied for supporting the choice of more sustainable energy solutions, mainly in the following areas: energy policy and management, evaluation of power generation technologies, evaluation of other energy systems, and electrical regional planning [20,21]. However, there is a lack in the existing literature concerning the use of MCDA tools for providing robust and sustainable choices within the bioenergy context [21]. According to the review conducted by Scott et al. [22], the most common topic studied in MCDA applied to bioenergy is the selection of the technology to use, comparing between technologies or between equipment within the same technology. The sustainability of bioenergy was only evaluated in 14% of the reviewed literature. This trend remains at present with only few studies addressing sustainable choices in the MCDA application to bioenergy systems (mainly application to biomass conversion routes [23,24]). In addition, little attention has been paid to MCDA studies of energy systems for heating in residential buildings [25–27]. These studies are mainly based on the use of economic and environmental indicators for the MCDA analysis.

This paper aims to fill the knowledge gap in the literature regarding the application of MCDA tools for bioenergy systems, as well as to provide recommendations for the selection of the best bioenergy options for residential heating in Southern Europe. Five different wood-based residential heating alternatives based on the work conducted by Quinteiro et al. [8] are evaluated in this study through the combination of three commercial technologies (pellet stove, wood stove and traditional fireplace) and the use of two different feedstocks widely available in the selected region (eucalypt and maritime pine species). In order to follow a sustainability perspective, the environmental analysis of the alternatives carried out in Quinteiro et al. [8] is complemented in this study with the evaluation of the economic and social dimensions for the prioritisation of alternatives, considering eight indicators (four environmental, two economic and two social). Moreover, an exploratory MCDA analysis is carried out in this work to obtain a synthesis of the results encompassing the three sustainability dimensions. Because of the wide availability of options concerning MCDA methods and weighting vectors, this exploratory analysis applies four of the most commonly used MCDA methods: Weighted Sum Method (WSM) [28], TOPSIS [29], ELECTRE [30], and PROMETHEE [31]. The sensitivity of the obtained rankings concerning the weighting approach is also carried out. Together, these analyses allow assessing the influence of the MCDA modelling options (method and weighting approach) on the recommended choice.

2. Materials and Methods

2.1. Five Wood-Based Residential Heating Alternatives

Five different wood-based residential heating alternatives for a single-family house are considered in this article to support decision-makers in developing sensible choices for the deployment of a European bioeconomy context. The three selected technologies represent the most widely used systems for residential heat supply in Southern Europe and more specifically in Portugal. In this sense, despite the traditional use of fireplaces, they are being replaced by wood stoves (for burning logs) and pellet stoves (for burning compressed biomass pellets) with higher energy conversion efficiency and lower air pollution. Additionally, wood pellets have higher energy density and thus require less space for storage than firewood. However, the selection of the most appropriate wood-fuelled system should be linked not only to technical indicators, but also economic, environmental and social criteria. Table 1 contains a brief description of the selected bioenergy options and the main assumptions taken into account for each one.

Table 1. List of wood-based alternatives residential heating considered in the multi-criteria decision analysis (MCDA) prioritisation.

Alternative	Code	Description [1]
Pellet stove, maritime pine	PS–MP	Thermal energy generation through combustion of maritime pine pellets (annual consumption of 0.53 t) in a pellet stove (energy conversion efficiency of 82%; 9.50 kW$_{th}$ of nominal power)
Wood stove, eucalypt	WS–E	Thermal energy generation through combustion of eucalyptus split logs (annual consumption of 0.70 t) in a wood stove (energy conversion efficiency of 65%; 18.20 kW$_{th}$ of nominal power)
Wood stove, maritime pine	WS–MP	Thermal energy generation through combustion of maritime pine split logs (annual consumption of 0.66 t) in a wood stove (energy conversion efficiency of 65%)
Fireplace, eucalypt	F–E	Thermal energy generation through combustion of eucalyptus split logs (annual consumption of 4.56 t) in a fireplace (energy conversion efficiency of 10%)
Fireplace, maritime pine	F–MP	Thermal energy generation through combustion of maritime pine split logs (annual consumption of 4.32 t) in a fireplace (energy conversion efficiency of 10%)

[1] The main assumptions taken into account for each alternative are based on Quinteiro et al. [8].

Figure 1 shows the life cycle stages of each alternative including: (i) forest management of both eucalypt and maritime pine species up to log loading onto trucks; (ii) pellets and wood split logs production; (iii) distribution of pellets and wood split logs; and (iv) thermal energy generation. The forest management stage includes all operations carried out during infrastructure establishment, site preparation, stand establishment, stand tending, and wood felling, forwarding and loading onto the truck. Regarding the production stage, five main operations are considered for the wood pellet alternative: (i) log chipping; (ii) milling; (iii) drying, using the heat produced from the combustion of maritime pine logging residues, chipped at the forest roadside; (iv) pelletising, and; and (v) packaging. In the case of split logs production, this process consists in splitting the wood logs into smaller portions of wood ready for burning.

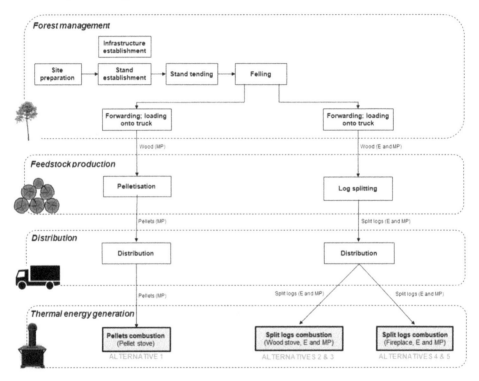

Figure 1. Simplified diagram of the wood-based residential heating alternatives in the MCDA prioritisation (E = eucalypt; MP = maritime pine).

Concerning the distribution stage of pellets, a distance of 15 km by truck is considered from pellets mill to the storage of wood pellets packed in LDPE bags (15 kg of pellets per bag). Additionally, another 15 km by private car from store to households is assumed. These average distances correspond to a 'short chain topology', involving proximity between the places of energy consumption and production [8]. In the case of the distribution of split logs, an average distance of 15 km by truck from the log splitting facilities to households has been considered [8].

Regarding the thermal energy generation stage, three wood-fuelled systems are selected, as previously mentioned, for residential heating based on the combustion of logs or pellets feedstock (from eucalypt or maritime pine): fireplace, wood stove and pellet stove. The pellet stove has a nominal thermal power of 9.5 kW$_{th}$ with an internal pellet storage tank and an auger screw to the burner for the provision of biomass feedstock. The wood stove presents a nominal power of 18.20 kW$_{th}$ and has a manual control of combustion air. Finally, the fireplace consists of a brick open fireplace operated manually. The annual consumption of feedstock for each alternative presented in Table 1 was calculated taking into consideration the requirement of fuel per MJ of heat [8] and the energy demand provided below. Further information regarding the features of the systems can be found in Quinteiro et al. [8].

The selected alternatives are used to meet the energy demand for space heating of the living room (total area 30 m^2) of a representative single-family house [32]. The house is located in the north of Portugal and the number of heating degree days considered in this region is 1652 °C·days, with an outdoor design temperature of 10 °C [32]. The estimation of the useful energy for heating is made following the methodology detailed in the decree-law n° 80/2006 of 4 April corresponding to the Portuguese Regulation of product characteristics for thermal performance in buildings [33]. Thus, 7978.85 MJ·yr^{-1} is the maximum useful energy required for the space heating of the living room of an average single-family house located in the north of Portugal according to the Equation (1) and the surface area:

$$N = 4.5 + (0.021 + 0.037 \cdot FF) \cdot GD \tag{1}$$

where *N* corresponds to the maximum value of useful energy required for heating (usually measured in kWh·m^{-2}·yr^{-1}); shape form (FF) is the ratio between the sum of the areas of the house envelope (outer and inner) and its respective interior volume; and GD is the number of degree days (i.e., a measure of the variation of 1 day's temperature against a standard reference temperature) of the area.

2.2. Criteria for Sustainability Assessment

The following eight indicators were selected for the sustainability assessment of wood-based alternatives: (i) global warming, (ii) terrestrial acidification, (iii) freshwater eutrophication, (iv) ozone formation (human health), (v) annualised capital costs, (vi) annual operation and maintenance costs, (vii) annual working hours for feedstock production, and (viii) annual number of days of absence due to non-fatal accidents during the feedstock production (Table 2). The sample of indicators are organised into a hierarchy that comprehends the three main dimensions of sustainability and their selection is motivated by the specific features of the case study. Furthermore, the chosen set of indicators represents commonly used indicators for evaluating the sustainability of energy systems [21], covering impacts for ecosystems and human health.

Table 2. Description of indicators included in the MCDA for prioritisation of wood-based residential heating alternatives.

Indicators	Code	Unit
Global warming potential	env1	kg CO$_2$ eq·yr^{-1}
Terrestrial acidification	env2	kg SO$_2$ eq·yr^{-1}
Freshwater eutrophication	env3	kg P eq·yr^{-1}
Ozone formation—human health	env4	kg NO$_x$ eq·yr^{-1}
Annualised capital costs	ec1	€·yr^{-1}
Annual operation and maintenance costs	ec2	€·yr^{-1}
Annual working hours—feedstock production	soc1	h·yr^{-1}
Annual number of days of absence due to non-fatal accidents	soc2	days·yr^{-1}

The environmental criteria were based on Quinteiro et al. [8], who calculated the indicators per 1 MJ of thermal energy generated by each alternative using LCA methodology. The suitability of life-cycle approaches as robust and reliable sources of environmental indicators is demonstrated in Martin-Gamboa et al. [21]. The functional unit is defined as 7978.85 MJ·yr^{-1} (i.e., the energy demand for space heating of the living room), as already explained. The environmental indicators were calculated with the ReCiPe 2016 impact assessment method [34].

The selected economic criteria encompass the annualised capital costs and the annual operation and maintenance costs. The first indicator includes costs of equipment and installation of the wood-based technologies. The average capital costs of these technologies range from approximately 1600 € in the case of pellet stoves to 1350 € (wood stoves) and 1050 € (fireplaces). These costs are estimated based on price information from suppliers and literature [35]. With the aim of providing the same temporal reference as the remaining indicators, capital costs are annualised multiplying them by the capital recovery factor (CRF), a ratio used to calculate the present value of an annuity [36]. According to the recommendations of García-Gusano et al. [37], an interest rate of 5% is used for the calculation of the annualised capital costs. This value agrees with other studies that establish the discount rate for households in energy system analysis between 3–5% [38]. Regarding the lifetime of the equipment under evaluation, 20 years is estimated for the pellet and wood stoves while a lifetime of 50 years is assumed for the fireplace, taking into account the average useful life of a residential house [32].

With respect to the second economic indicator (annual operation and maintenance costs), it includes cost of feedstock, electricity, maintenance, and contingencies (e.g., breakdowns). This indicator is computed according to the heat demand of 7978.85 MJ·yr^{-1}. Based on the Association of Forestry Producers of Portugal [39], a price of 0.056 €·kg^{-1} is estimated for eucalypt wood and 0.046 €·kg^{-1} for maritime pine. In the case of pellet feedstock, an average price of 3.35 € per bag of 15 kg is assumed according to direct suppliers. An electricity price of 0.23 €·kWh^{-1} is used, which is obtained from the statistics of the European Union for Portugal [40]. Lastly, maintenance costs are calculated taking into account a percentage (4%) of capital costs, while contingency costs are assumed as 10% of the operation and maintenance costs [41].

Finally, the social criteria are the annual working hours and the annual number of days of absence due to non-fatal work-related accidents. Both indicators cover the stages needed to produce the feedstock (i.e., forest management and biomass feedstock production stages). The information for the calculation of the annual working hours were obtained from Dias and Arroja [6] and Dias et al. [42], used for obtaining the working hours in terms of forest management, as well as from Hunsberger and Mosey [43], used for providing the working hours required for the whole pellet production process. The working hours linked to the logs splitting were obtained from direct communication. It is important to keep in mind that the calculated working hours refer to the effective working time required to provide the annual amount of feedstock for each of the wood-based alternatives. Additionally, they can also be understood as a contribution of the forestry sector to local economic development.

The number of days of absence is calculated according to statistical data from Dos Santos [44] for the year 2014 (last year for which data are available). Firstly, the number of non-fatal work-related accidents (per worker) and the days of absence due to these accidents (per year) in the forestry sector for some districts of the north region of Portugal were obtained (see Tables A1 and A2 of the Appendix A). The following districts were considered: Aveiro, Braga, Coimbra, Leiria, Porto, Viana do Castelo, Vila Real, and Viseu. These districts were selected because they represent the largest area of distribution of eucalypt and maritime pine forests in Portugal [45]. From the statistical data of work-related accidents and their corresponding days of absence, the average values of the whole region considered were obtained and subsequently linked to the working hours previously calculated to obtain the annual days of absence specific to each alternative under evaluation.

2.3. MCDA Methods

The choice of the best option in terms of sustainability is a difficult decision that usually entails the application of an MCDA method. Within the wide range of MCDA methods currently available, the selection of a specific one highly depends on the particular characteristics of each problem (e.g., identification of the most suitable energy technology for satisfying local electricity demand) and the decision-makers' needs. To ensure the robustness of the present analysis, four MCDA techniques are applied in this case study, viz., WSM, TOPSIS, ELECTRE, and PROMETHEE. The rationale behind the selection of these tools is that they are widely applied in the literature [12,17,21] and follow different principles.

2.3.1. Weighted Sum Method

The WSM [28] follows the principle of aggregating the performances of the alternatives on the multiple criteria according to an additive aggregation:

$$S(a_i) = w_1 \cdot v_1(a_i) + \dots + w_n \cdot v_n(a_i) \tag{2}$$

where $S(a_i)$ denotes the global value of the i-th alternative, ai; n denotes the number of criteria (indicators); w_j denotes the weight of the j-th criterion; and $v_n(a_i)$ denotes the value of ai on the j-th criterion ($j = 1, \dots, n$). Since the impacts are measured in different mathematical scales, the values aggregated in Equation (2) must be transformed so that scales are commensurate. This can be accomplished by a normalisation operation, with the following being one of the most popular and selected in the present work to transform the impact of an alternative ai according to the j-th criterion, denoted $I_j(a_i)$:

$$v_j(a_i) = |I_j(worst_j) - I_j(a_i)|/|I_j(worst_j) - I_j(best_j)| \tag{3}$$

where $worst_j$ and $best_j$ denote the alternatives with the worst and best potential sustainability impact according to the j-th indicator. According to Equation (3), the normalised values on each indicator vary between 0 for the worst alternative to 1 for the best alternative.

It should be noted that the previous normalisation, despite being very popular, can be biased by the introduction of a new alternative that is extremely good or extremely bad on some indicator [46], and so it might be preferable to use a normalisation based on a fixed reference [47,48]. Another sensible option is to replace normalisation by building value functions that reflect the preferences of a decision-maker [18,19].

2.3.2. TOPSIS Method

The TOPSIS tool [29] follows the principle of aggregating the performances of the alternatives on the multiple criteria into a single value representing how close it is to an ideal solution, represented by the vector ($best_1, \dots, best_n$), and how far it is from an anti-ideal solution, represented by the vector ($worst_1, \dots, worst_n$). Denoting by $d_i^+(a_i)$ the Euclidean distance between an alternative a_i and the ideal solution and denoting by $d_i^-(a_i)$ the Euclidean distance between an alternative a_i and the anti-ideal solution, the score of a_i is given by:

$$S(a_i) = d_i^-(a_i)/[d_i^-(a_i) + d_i^+(a_i)] \tag{4}$$

Thus, the score varies between 0 and 1 (and more is better, meaning that the alternative is farther away from the anti-ideal). As in the WSM, before computing distances, the mathematical scales must be normalised so that scales are commensurate. In TOPSIS, normalisation consists in dividing the impacts on each criterion by the square root of the sum of the squares of the impacts of the alternatives being compared. These normalised values are also multiplied by the weights of the criteria, so that the Euclidean distances used in Equation (4) are weighted distances. It should be noted that the weights are non-negative and their sum is equal to 1.

2.3.3. ELECTRE Method

The ELECTRE method [30] follows the principle of building an outranking relation S resulting from comparisons between each alternative and its competitors. In the original ELECTRE variant, the outranking relation is crisp (true/false) rather than valued (values in [0,1]), requiring the definition of a concordance threshold (required majority). When comparing two alternatives, a_i and a_j, ELECTRE adds up the total weight of the criteria on which a_i is at least as good as a_j (concordance) and checks whether a_i is worse than a_j by a large difference on any criterion (discordance). Then, based on the concordance threshold c (required majority) and the discordance threshold d (maximum discordance) it concludes if globally a_i is at least as good as (outranks) a_j, denoted $a_i \, S \, a_j$:

$$a_i \, S \, a_j \Leftrightarrow \text{Sum}\{\text{for } j: \, v_j(a_i) \geq v_j(a_j)\}w_j \geq c \wedge \text{Max}\{\text{for } j: \, v_j(a_i) < v_j(a_j)\}[v_j(a_j) - v_j(a_i)] \leq d \qquad (5)$$

The value of the alternatives on a given criterion, $v_j(a_i)$ and $v_j(a_j)$, needs to be normalised according to Equation (3). In this work, the concordance threshold c was set to its minimum value 0.51 (a simple majority), and the discordance threshold d was set to 0.8 (discordance vetoes an outranking relation if it is greater than 80% of the scale amplitude).

To obtain comparable results with the WSM and TOPSIS, which yield a ranking of the alternatives, there are several variants of ELECTRE available [49]. One of the simplest ways of obtaining a ranking, which has a clear axiomatic foundation, is the net flow approach and its particular instance of the Copeland method [50]. According to this method, alternatives are ranked by the number of "wins" (alternatives they outrank) minus the number of "losses" (alternatives that outrank them). The ELECTRE score used in this work for ranking the alternatives in a set A is then:

$$S(a_i) = \#\{a_j \in A: a_i \, S \, a_j\} - \#\{a_j \in A: a_j \, S \, a_i\} \qquad (6)$$

2.3.4. PROMETHEE Method

The PROMETHEE method [31] also follows the principle of building an outranking relation resulting from comparisons between each alternative and its competitors. However, the outranking relation is always valued (values in [0,1]), it computes only the concordance that one alternative is preferred to others (not computing discordance opposed by the minority criteria), and it does not rely on the notion of a required majority threshold.

Similarly to ELECTRE, the PROMETHEE method compares each alternative against all the competing alternatives in a set A. The preference degree of an alternative a_i over another a_j is equal to 0 if a_i is worse than a_j, is equal to 1 if a_i is better than a_j by a difference of p_j or larger (p_j is a parameter named preference threshold), and is a value between 0 and 1 obtained by linear interpolation if the advantage of a_i over a_j is positive but less than p_j. This way of computing a preference index corresponds to Type III, one of the six formulae available in PROMETHEE to compute it [31]. In this work, on each criterion, p_j was considered to be 50% of the average performance of the alternatives compared. Then, the single-criterion preference indexes are aggregated as a weighted sum taking into account the criteria weights, to obtain the overall preference of a_i over a_j, denoted $\pi(a_i, a_j)$.

Finally, the ranking of the alternatives in PROMETHEE follows a net flow approach, based on the following Φ scores:

$$\Phi(a_i) = \text{Sum}\{\text{for } a_j \in A\}[\pi(a_i, a_j) - \pi(a_j, a_i)] \qquad (7)$$

3. Results and Discussion

In the present section, firstly the sustainability profiles of each wood-based system are presented. Then, the alternatives are prioritised through the four MCDA methods selected: WSM, TOPSIS, ELECTRE, and PROMETHEE. Finally, a sensitivity analysis of the rankings obtained with regard to the weighting approach is carried out. In this respect, the weights are modified—each "weighting scenario" emphasizes a different dimension of sustainability—to evaluate their influence in the original ranking (where an equal-weighted approach is taken).

3.1. Sustainability Assessment of Wood-Based Alternatives

Table 3 contains the results obtained for each indicator of the sustainability assessment. The highest impact values for the environmental criteria are associated with the alternatives based on the use of a fireplace. In particular, the alternative based on eucalypt feedstock (F–E) presents the greatest impacts in all the environmental indicators considered, negatively affecting its sustainability performance. This unfavourable performance is strongly linked to an inefficient combustion of wood split logs, which requires a higher amount of feedstock and involves a higher emission of gaseous pollutants per MJ of thermal energy produced in comparison with the rest of alternatives. Regarding the identification of the best alternative in terms of environmental criteria there is not a clear trend, with the pellet stove alternative (PS–MP) being the most favourable in global warming (env1) and ozone formation (human health) (env4), while wood stove (eucalypt, WS–E and maritime pine, WS–MP) performs better in terms of terrestrial acidification (env2) and freshwater eutrophication (env3).

Table 3. Indicator results applied in the MCDA for prioritisation of wood-based residential heating alternatives.

Alternative	Env1	Env2	Env3	Env4	Ec1	Ec2	Soc1	Soc2
PS–MP	62.24	0.63	$1.49 \cdot 10^{-2}$	1.47	126.62	210.34	1.92	$1.39 \cdot 10^{-3}$
WS–E	181.12	0.64	$4.72 \cdot 10^{-3}$	2.31	108.10	107.27	1.31	$9.85 \cdot 10^{-4}$
WS–MP	177.93	0.47	$4.01 \cdot 10^{-3}$	2.14	108.10	97.46	1.18	$8.92 \cdot 10^{-4}$
F–E	973.42	4.09	$1.84 \cdot 10^{-2}$	14.84	57.48	328.64	8.49	$6.40 \cdot 10^{-3}$
F–MP	957.46	2.97	$1.40 \cdot 10^{-2}$	13.80	57.48	265.16	7.69	$5.80 \cdot 10^{-3}$

In the case of economic criteria, it is not possible to identify a best or worst alternative. Despite presenting the lowest installation costs, PS–MP is the worst alternative in terms of annualised capital costs (ec1) due to its high market price in comparison with the remaining alternatives (more than 1300 € on average). On the other hand, operation and maintenance costs (ec2) are strongly affected by the annual feedstock costs. Thus, the systems based on the use of a fireplace to meet the residential heat demand (especially in the case of F–E with higher prices of the biomass feedstock) are identified as the worst alternatives because they require the largest amounts of biomass fuel within the set of options evaluated.

Finally, a similar not-dominant trend is found in the case of social criteria. In contrast to the remaining indicators, the annual working hours (soc1) should be maximised with the aim of contributing to the local economic development. Fireplace alternatives are the most favourable systems in terms of annual working hours (mainly F–E) due to the longer working times employed in these alternatives to provide the annual amount of biomass feedstock (more than 4 tonnes of eucalypt or maritime pine). For this indicator, wood stove alternatives emerge as the worst systems with the lowest working hours (especially WS–MP). It should be noted that longer working times are associated with a higher probability of accidents [44]. For this reason, fireplace alternatives are identified as the worst alternatives when days of absence due to non-fatal work accidents are estimated. The observed variability of indicators results makes it difficult to select directly the most sustainable bioenergy systems. Therefore, the use of MCDA tools emerges as necessary and convenient to prioritise the set of wood-based alternatives according to the multiple criteria evaluated.

3.2. MCDA Prioritisation of Wood-Based Alternatives

The data presented in Table 3 constitute the matrix which is the input for an own-developed MCDA model implemented in Excel to compute the scores. The results obtained with the four different tools (viz., WSM, TOPSIS, ELECTRE, and PROMETHEE) applying the same weights for all the indicators are presented in Table 4. The scores obtained have different scales due to the specific mathematical formulation of each method (Section 2.3).

Table 4. Scores (equal weighted) of wood-based residential heating alternatives obtained through application of: (i) WSM, (ii) TOPSIS, (iii) ELECTRE, and (iv) PROMETHEE.

Alternative	WSM	TOPSIS	ELECTRE	PROMETHEE
PS–MP	0.590	0.623	−1	0.162
WS–E	0.742	0.678	−1	0.425
WS–MP	0.762	0.682	2	0.475
F–E	0.250	0.321	−1	−0.662
F–MP	0.374	0.364	1	−0.400

Differences in the ranking of alternatives due to the application of the different MCDA tools can be observed in Figure 2. In this sense, the best alternative according to their position in the four rankings is WS–MP. This alternative presents the highest scores calculated through the application of the four MCDA tools included in the analysis. Despite this fact, the scores are similar between the two wood stoves alternatives (with exception of the application of ELECTRE method). Thus, both WS–MP and WS–E could be considered as appropriate bioenergy systems for residential heating in terms of sustainability performance. Regarding the worst alternative, F–E is ranked last in all rankings (in the case of ELECTRE, tied in the last position with the wood stove and fireplace alternatives based on the combustion of eucalypt split logs).

The similarity between rankings shows that the results are quite robust relative to the choice of the MCDA tools used for the present case study. The only differences are found when the ELECTRE method is applied for the prioritisation of the alternatives. These differences stem from the strong non-compensatory nature of the discordance condition in ELECTRE. Since the pellet and wood stove options (PS–MP, WS–E, and WS–MP) are much worse than the fireplace options (F–E and F–MP) on the indicators annualised capital costs (ec1) and annual working hours (soc1), none of the former three alternatives can outrank the latter two. On the other hand, since the fireplace options F–E and F–MP are so much worse on the remaining indicators, neither can outrank PS–MP, WS–E or WS–MP. These form two incomparable groups, according to ELECTRE. In the first group, PS–MP and WS–E are outranked by WS–MP and outrank no other alternative; hence they are placed in the third (viz., last) position with a net flow of -1. On the other hand, F–MP benefits because it is capable of outranking the other fireplace alternative F–E while no other alternative outranks it.

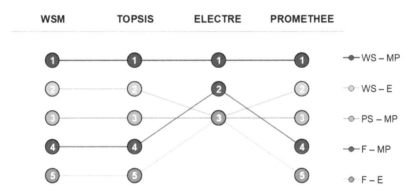

Figure 2. Ranking (equal weighted) of wood-based residential heating alternatives obtained through application of: (i) WSM, (ii) TOPSIS, (iii) ELECTRE, and (iv) PROMETHEE.

3.3. Influence of Weighting

The results presented in the previous section correspond to considering equal weights for all the indicators. However, in MCDA, different decision-makers will have other preferences. Depending on the perspective, they might wish to emphasise environmental, economic, or social criteria, possibly leading to different results than those obtained by considering equal weighting. It is therefore important to analyse how sensitive the results are to changes in the sample of weights.

A possible way of assessing the effect of using different weight vectors is to consider different "weighting scenarios", each one overweighting a given sustainability dimension. Table 5 presents an example of this strategy, placing 75% of the total weight on a single dimension at a time (i.e., the weight of each dimension is the triple of the other two dimensions together). This is a common strategy to represent the preference of a decision-maker in a specific dimension. Computing the results for these weights using the same methods as before does not lead to major changes. In fact, the rankings presented in Figure 2 are maintained in the majority of the cases. When the weights prioritising the environmental dimension are used, all the rankings remain unchanged except in the case of ELECTRE, placing the WS–E alternative in the last position. When the weights prioritising the economic dimension are used, the ranking changes for the WSM, ELECTRE and PROMETHEE, with PS–MP falling from 3rd to 5th place (F–MP and F–E therefore climb one position). When the weights prioritising the social dimension are used, the ranking changes only for the TOPSIS method, with PS–MP climbing from 3rd to 1st place (WS–MP and WS–E therefore drop one position to become 2nd and 3rd). In the latter case, however, the scores of PS–M become very close to the scores of WS–MP and WS–E.

Table 5. Weight vectors emphasising each sustainability dimension with 75% of the weight.

Priority	w_{Env1}	w_{Env2}	w_{Env3}	w_{Env4}	w_{Ec1}	w_{Ec2}	w_{Soc1}	w_{Soc2}
Environmental	0.1875 (3/16)	0.1875 (3/16)	0.1875 (3/16)	0.1875 (3/16)	0.0625 (1/16)	0.0625 (1/16)	0.0625 (1/16)	0.0625 (1/16)
Economic	0.0417 (1/24)	0.0417 (1/24)	0.0417 (1/24)	0.0417 (1/24)	0.375 (3/8)	0.375 (3/8)	0.0417 (1/24)	0.0417 (1/24)
Social	0.0417 (1/24)	0.0417 (1/24)	0.0417 (1/24)	0.0417 (1/24)	0.0417 (1/24)	0.0417 (1/24)	0.375 (3/8)	0.375 (3/8)

Rather than selecting a few different weighting vectors as described above, it is also possible to perform a stochastic analysis simulating thousands of weight vectors drawn randomly from a uniform distribution [48,51]. Using Monte-Carlo simulation, it is then possible to obtain statistics about the results corresponding to these stochastic weights.

Table 6 shows the Monte-Carlo simulation results (obtained with the @Risk software) for 50,000 random weight vectors uniformly distributed on the simplex defined by the constraints that weights are non-negative and their sum is equal to 1, following the process of Butler et al. [52]. Since ELECTRE yields tie into the rankings frequently (e.g., Figure 2), in this stochastic analysis the number of "losses" was used as a tie-breaking procedure (but some ties still remain when the number of "losses" is the same among tied alternatives).

It can be observed that the rankings presented in Figure 2 are rather stable, but in extreme cases they can be quite different. Just as an example, according to the WSM, PS–MP is ranked 3rd for 79.2% of the random weights, but for weights such as, e.g., $w = (0.21, 0.02, 0.01, 0.36, 0.01, 0.03, 0.35, \text{and } 0.01)$ it is ranked 1st and for weights such as, e.g., $w = (0.15, 0.02, 0.22, 0.03, 0.01, 0.01, 0.49, \text{and } 0.07)$ it is ranked 5th. The probability of obtaining the same rankings of Figure 2 is higher for the WSM and PROMETHEE, but TOPSIS and ELECTRE also establish the ranking positions in Figure 2 with high probability.

Table 6. Rank probabilities with stochastic weights.

	Alternative	Rank 1st	Rank 2nd	Rank 3rd	Rank 4th	Rank 5th
	PS–MP	1.5%	1.0%	79.2%	7.5%	10.8%
	WS–E	-	90.4%	5.1%	4.4%	0.1%
WSM	WS–MP	91.8%	3.6%	4.0%	0.4%	0.2%
	F–E	0.5%	4.0%	0.3%	6.4%	88.9%
	F–MP	6.2%	1.0%	11.5%	81.3%	<0.1%
	PS–MP	11.7%	1.6%	73.6%	3.0%	10.1%
	WS–E	<0.1	77.0%	11.9%	8.1%	3.0%
TOPSIS	WS–MP	76.3%	10.5%	9.3%	2.8%	1.1%
	F–E	2.1%	8.6%	0.3%	6.5%	82.6%
	F–MP	9.9%	2.4%	5.0%	79.6%	3.2%
	PS–MP	24.1%	0.5%	27.3%	-	48.2%
	WS–E	2.8%	0.2%	49.9%	1.4%	45.8%
ELECTRE	WS–MP	78.8%	0.2%	19.0%	1.2%	0.8%
	F–E	-	2.9%	6.6%	90.6%	-
	F–MP	3.1%	93.4%	2.8%	0.8%	-
	PS–MP	5.4%	3.2%	74.4%	8.3%	8.7%
	WS–E	-	84.0%	11.3%	4.2%	0.5%
PROMETHEE	WS–MP	87.6%	7.5%	3.4%	0.9%	0.5%
	F–E	0.5%	4.1%	0.4%	4.7%	90.3%
	F–MP	6.5%	1.2%	10.5%	81.9%	<0.1%

3.4. Influence of Hierarchy

The analysis reported so far considers all the sustainability dimensions on an equal standing, but this ignores a potential issue: The environmental dimension is represented by four criteria, whereas the economic dimension and the social dimension are represented by only two criteria each. This means that the results presented in Section 3.2, considering equal weights, are in fact overweighting the environmental dimension.

If decision-makers wish that all the sustainability dimensions have the same weight (1/3), then a different weighting vector should be applied: dividing 1/3 by four yields a weight of 1/12 for each one of the four environmental indicators; dividing 1/3 by two yields a weight of 1/6 for each one of the two economic indicators, and the same weights apply to the two social indicators. Given this weighting vector, the scores in Table 4 would be changed to those in Table 7. Nevertheless, the ranking of the alternatives (Figure 2) would remain the same except in the case of ELECTRE, with PS–MP falling from 3rd to 5th place (F–E drop one position to become 4th).

Table 7. Scores (equal weighted considering hierarchy) of wood-based residential heating alternatives obtained through application of: (i) WSM, (ii) TOPSIS, (iii) ELECTRE, and (iv) PROMETHEE.

Alternative	WSM	TOPSIS	ELECTRE	PROMETHEE
PS–MP	0.520	0.541	−2	−0.015
WS–E	0.681	0.594	0	0.217
WS–MP	0.697	0.598	2	0.247
F–E	0.333	0.411	−1	−0.314
F–MP	0.439	0.444	1	−0.135

The same consideration is in order if a stochastic weights analysis is performed. Indeed, since there are more environmental criteria than economic or social ones, the environment dimension will on average have more weight than the economic or the social dimension. Correcting this requires a two-step strategy to generate random weights. In a first step, the total weight of each dimension $(w_{Env}, w_{Ec}, w_{Soc})$ is generated randomly such that $(w_{Env}, w_{Ec}, w_{Soc}) \geq 0$ and $w_{Env} + w_{Ec} + w_{Soc} = 1$. Then, the relative sub-weights of the four environmental indicators $(s_{env1}, s_{env2}, s_{env3}, s_{env4})$ are randomly generated such that $(s_{env1}, s_{env2}, s_{env3}, s_{env4}) \geq 0$ and $s_{env1} + s_{env2} + s_{env3} + s_{env4} = 1$. Setting the sub-weights for the economic indicators requires only one random value, s_{ec1}, drawn from the uniform distribution in [0,1], and then setting $s_{ec2} = 1 - s_{ec1}$. Similarly, a random value, s_{soc1} is generated, allowing to define $s_{soc2} = 1 - s_{soc1}$. Finally, dimension weights and sub-weights are combined so that $w_{Env1} = w_{Env} \cdot s_{env1}, \ldots, w_{Env4} = w_{Env} \cdot s_{env4}, w_{Ec1} = w_{Ec} \cdot s_{ec1}, w_{Ec2} = w_{Ec} \cdot s_{ec2}, w_{Soc1} = w_{Soc} \cdot s_{soc1}$, and $w_{Soc2} = w_{Soc} \cdot s_{soc2}$.

Considering the stochastic weights as described above, the probability results in Table 6 are replaced by the probability results in Table 8. In every case, the most likely rank has a probability greater than about 60% and these most likely rankings still coincide with Figure 2. However, by comparing Table 6 with Table 8 one can observe that the latter rankings are less stable in all the cases, i.e., the most likely ranking has less probability. The reason for this is that respecting the hierarchical structure increases the role of the economic and social indicators in the final result, and indicators ec1 and soc1 do not favour the top-ranked alternatives. Since now they have more chances of influencing the result, the scores of the top-ranked three alternatives tend to decrease and so they can more often be surpassed by other alternatives.

Table 8. Rank probabilities with stochastic weights considering the dimension-indicators hierarchy.

	Alternative	Rank 1st	Rank 2nd	Rank 3rd	Rank 4th	Rank 5th
WSM	PS–MP	1.4%	0.9%	61.7%	8.8%	27.2%
	WS–E	-	74.8%	9.0%	15.8%	0.3%
	WS–MP	76.5%	5.2%	13.5%	2.8%	2.0%
	F–E	4.5%	13.7%	0.5%	10.7%	70.5%
	F–MP	17.6%	5.3%	15.2%	61.8%	<0.1%
TOPSIS	PS–MP	7.3%	1.0%	65.7%	3.6%	22.4%
	WS–E	<0.1%	65.9%	9.1%	20.0%	5.0%
	WS–MP	65.7%	7.9%	18.2%	5.1%	3.2%
	F–E	7.1%	17.8%	0.3%	7.0%	67.8%
	F–MP	19.8%	7.4%	6.8%	64.3%	1.7%
ELECTRE	PS–MP	21.0%	0.5%	14.7%	-	63.7%
	WS–E	11.9%	0.5%	59.7%	1.7%	26.2%
	WS–MP	80.5%	0.5%	11.5%	1.6%	5.9%
	F–E	0.1%	8.8%	14.2%	77.0%	-
	F–MP	1.8%	80.4%	11.9%	6.0%	-
PROMETHEE	PS–MP	3.8%	2.1%	60.7%	9.8%	23.5%
	WS–E	<0.1%	71.4%	13.6%	14.3%	0.7%
	WS–MP	74.1%	7.9%	10.4%	4.0%	3.6%
	F–E	4.6%	13.1%	0.7%	9.5%	72.2%
	F–MP	17.5%	5.5%	14.6%	62.4%	<0.1%

Even though TOPSIS places WS–MP in the first position with a relatively low probability of 65.7% (Table 8), the analysis carried out in the present section corroborates the results from Figure 2. Overall, the main conclusion that WS–MP is the most preferred option for all the methods is still fairly robust.

3.5. Final Remarks

The current development of energy policies, roadmaps or plans at all levels (i.e., national, regional, local, etc.) should be in line with the Sustainable Development Goals (SDGs) agreed by the United Nations [53]. The exploratory MCDA analysis conducted in the present study for the sustainability-oriented prioritisation of wood-based bioenergy systems can contribute to set strategies aligned with those SDGs. In particular, it facilitates the decision-making process and the provision of plans that address most of the targets covered by both SDG7 (access to affordable and clean energy) and SDG13 (climate change mitigation). Additionally, links with some of the targets related to well-being (SDG3), decent work and economic growth (SDG8), and sustainable regions (SDG11) could also be identified if the outcomes of this study are taken into account by decision-makers in the elaboration of plans. It is important to note that the relevance of the outcomes of this research is not limited to energy actors in Portugal, but they are relevant to any decision-maker considering the development of energy plans, roadmaps or strategies under sustainability aspects.

The significant involvement of stakeholders (e.g., policy-makers, actors from biomass-related sector, local community or consumers) usually found in the energy sector encompasses a wide range of views which could significantly vary the outcomes of the prioritisation. The weighting scenarios reported in this study allow visualising the influence of different actors' preferences (i.e., criteria weights) leading to contrasting and practical recommendations. Regarding the choice of indicators (e.g., type and number per sustainability dimension) and technical options within the MCDA methods (e.g., normalisation, concordance and discordance thresholds, etc.), special attention should be paid to the selection of these elements due to their influence on the final results. In this study, indicators and technical options were chosen by academics, but a future involvement of decision-makers at company or policy level may have modified those elements. Thus, a sensitivity analysis on the choice of indicators and technical MCDA options should be further studied. In addition, the incorporation of a life-cycle perspective into the economic and social criteria is also a future direction to explore.

4. Conclusions

An MCDA exploratory analysis was carried out to prioritise five bioenergy alternatives for residential heating in Southern Europe filling a gap in the MCDA application to bioenergy systems. The highest impacts for the environmental criteria were identified for the fireplace due to inefficient combustion of wood split logs, while there was not a preferred environmental performance option. Regarding the economic and social dimensions, no clear trend was found as the alternatives present favourable or unfavourable performances depending on the criterion evaluated. This fact strongly supported the need to apply MCDA methods for the prioritisation of alternatives in terms of sustainability.

The MCDA prioritisation revealed the wood stove options (similar scores for combustion of maritime pine or eucalypt logs) as the most appropriate alternatives in terms of sustainability performance. Regarding the worst alternative, fireplaces based on the combustion of eucalypt logs were ranked last in all MCDA rankings. Furthermore, analyses in terms of how sensitive were the results to changes in weighting were carried out in order to incorporate the vision of decision-makers and to facilitate their role in the prioritisation of bioenergy alternatives for residential heating. These analyses confirmed wood stoves with the combustion of maritime pine logs as the leading alternative and the key role of the analysts within this type of MCDA studies. Therefore, an appropriate prioritisation of bioenergy alternatives in terms of sustainability was demonstrated to be relevant for paving the way to a future wide deployment of the European bioeconomy sector, being the role of the analyst key to provide robust and reliable decisions.

Author Contributions: M.M.-G. conceived the study, calculated the economic and social indicators and implemented the MCDA methods; L.C.D. implemented the MCDA methods and developed the stochastic analysis; P.Q. calculated and provided the environmental indicators; all authors analysed the data and contributed to writing the paper.

Funding: "This research was partly funded by CESAM (UID/AMB/50017/2019), by FEDER (project SABIOS code PTDC/AAGMAA/6234/2014 - POCI-01-0145-FEDER-016765), and by FCT (contracts CEECIND/00143/2017 and CEECIND/02174/2017)".

Acknowledgments: Martín-Gamboa states that thanks are due to FCT/MCTES for the financial support to CESAM (UID/AMB/50017/2019), through national funds. This work is a contribution to the project SABIOS (PTDC/AAGMAA/6234/2014) funded under the project 3599-PPCDT by FEDER, through COMPETE2020 - Programa Operacional Competitividade e Internacionalização (POCI), and by national funds, through FCT/MCTES. Thanks are also given to the FCT for contracts granted to Paula Quinteiro (CEECIND/00143/2017) and to Ana Cláudia Dias (CEECIND/02174/2017).

Conflicts of Interest: The authors declare no conflict of interest. The funders had no role in the design of the study; in the collection, analyses, or interpretation of data; in the writing of the manuscript, or in the decision to publish the results.

Appendix A

Tables A1 and A2 present the number of non-fatal work-related accidents (per worker) and the annual days of absence due to these accidents in the forestry sector for the selected districts of the north region of Portugal. These values were based on statistical data of agriculture, forestry and fishing sectors from Dos Santos [44]. Within this estimation, a simplifying assumption is made such that data from these sectors was totally allocated to forestry sector due to lack of availability of local information.

Table A1. Non-fatal work-related accidents per worker.

Districts	Non-Fatal Work-Related Accidents
Aveiro	4.50×10^{-2}
Braga	3.14×10^{-2}
Coimbra	5.17×10^{-2}
Leiria	7.31×10^{-2}
Porto	3.57×10^{-2}
Viana do Castelo	2.14×10^{-2}
Vila real	1.43×10^{-2}
Viseu	3.56×10^{-2}

Table A2. Annual days of absence due to non-fatal work-related accidents.

Districts	Annual Days of Absence
Aveiro	48.12
Braga	34.95
Coimbra	40.00
Leiria	32.14
Porto	34.72
Viana do Castelo	40.94
Vila real	60.09
Viseu	34.64

References

1. Dahiya, S.; Kumar, A.N.; Sravan, J.S.; Chatterje, S.; Sarkar, O.; Mohan, S.V. Food waste biorefinery: Sustainable strategy for circular bioeconomy. *Bioresour. Technol.* **2018**, *248*, 2–12. [CrossRef] [PubMed]
2. Patterman, C.; Aguilar, A. The origins of the bioeconomy in the European Union. *New Biotechnol.* **2018**, *40*, 20–24. [CrossRef] [PubMed]
3. European Commission. *Communication from the Commission to the European Parliament, the Council, the European Economic and Social Committee and the Committee of the Regions. Innovating for Sustainable Growth: A Bioeconomy for Europe*; EC: Brussels, Belgium, 2012.
4. Ronzon, T.; M'Barek, R. *Brief on Jobs and Growth of the Bioeconomy 2009–2015*; Publications Office of the European Union: Luxembourg, 2018.
5. Phillips, D.; Mitchell, E.J.S.; Lea-Langton, A.R.; Parmar, K.R.; Jones, J.M.; Williams, A. The use of conservation biomass feedstocks as potential bioenergy resources in the United Kingdom. *Bioresour. Technol.* **2016**, *212*, 271–279. [CrossRef] [PubMed]
6. Dias, A.C.; Arroja, L. Environmental impacts of eucalypt and maritime pine wood production in Portugal. *J. Clean. Prod.* **2012**, *37*, 368–376. [CrossRef]
7. Nunes, J.; Freitas, H. An indicator to assess the pellet production per forest area. A case-study from Portugal. *For. Policy Econ.* **2016**, *70*, 99–105. [CrossRef]
8. Quinteiro, P.; Tarelho, L.; Marques, P.; Martin-Gamboa, M.; Freire, F.; Arroja, L.; Dias, A.C. Life cycle assessment of wood pellets and wood split logs for residential heating. *Sci. Total Environ.* **2019**, *689*, 580–589. [CrossRef]
9. Strzalka, R.; Schneider, D.; Eicker, U. Current status of bioenergy technologies in Germany. *Renew. Sustain. Energy Rev.* **2017**, *72*, 801–820. [CrossRef]

10. Ferreira, J.; Esteves, B.; Cruz-Lopes, L.; Evtuguin, D.V.; Domingos, I. Environmental advantages through producing energy from grape stalk pellets instead of wood pellets and other sources. *Int. J. Environ. Stud.* **2018**, *75*, 812–826. [CrossRef]

11. Paredes-Sánchez, J.P.; López-Ochoa, L.M.; López-González, L.M.; Las-Heras-Casas, J.; Xiberta-Bernata, J. Evolution and perspectives of the bioenergy applications in Spain. *J. Clean. Prod.* **2019**, *213*, 553–568. [CrossRef]

12. Zhou, P.; Ang, B.W.; Poh, K.L. Decision analysis in energy and environmental modeling: An update. *Energy* **2006**, *31*, 2604–2622. [CrossRef]

13. International Organization for Standardization. *ISO 14040:2006 Environmental Management—LIFE Cycle Assessment—Principles and Framework*; ISO: Geneva, Switzerland, 2006.

14. International Organization for Standardization. *ISO 14044:2006 Environmental Management—Life Cycle Assessment—Requirements and Guidelines*; ISO: Geneva, Switzerland, 2006.

15. Munda, G. Multi criteria decision analysis and sustainable development. In *Multiple Criteria Decision Analysis: State of the Art Surveys*; Figueira, J., Greco, S., Ehrogott, M., Eds.; Springer: New York, NY, USA, 2005; pp. 953–986.

16. Wang, J.J.; Jing, Y.Y.; Zhang, C.F.; Zhao, J.H. Review on multi-criteria decision analysis aid in sustainable energy decision-making. *Renew. Sustain. Energy Rev.* **2009**, *13*, 2263–2278. [CrossRef]

17. Ibáñez-Forés, V.; Bovea, M.D.; Pérez-Belis, V. A holistic review of applied methodologies for assessing and selecting the optimal technological alternative from a sustainability perspective. *J. Clean. Prod.* **2014**, *70*, 259–281. [CrossRef]

18. Belton, V.; Stewart, T. *Multiple Criteria Decision Analysis. An Integrated Approach*; Kluwer: Boston, MA, USA, 2002.

19. Dias, L.C.; Silva, S.; Alçada-Almeida, L. Multi-criteria environmental sustainability assessment with an additive model. In *Handbook on Methods and Applications in Environmental Studies*; Ruth, M., Ed.; Edward Elgar: Northampton, UK, 2015; pp. 450–472.

20. Antunes, C.H.; Henriques, C.O. Multi-Objective Optimization and Multi-Criteria Analysis Models and Methods for Problems in the Energy Sector. In *Multiple Criteria Decision Analysis*; Greco, S., Ehrgott, M., Figueira, J., Eds.; Springer: New York, NY, USA, 2016.

21. Martín-Gamboa, M.; Iribarren, D.; García-Gusano, D.; Dufour, J. A review of life-cycle approaches coupled with Data Envelopment Analysis within multi-criteria decision analysis for sustainability assessment of energy systems. *J. Clean. Prod.* **2017**, *150*, 164–174. [CrossRef]

22. Scott, J.A.; Ho, W.; Dey, P.K. A review of multi-criteria decision-making methods for bioenergy systems. *Energy* **2012**, *42*, 146–156. [CrossRef]

23. Schröder, T.; Lauven, L.P.; Beyer, B.; Lerche, N.; Geldermann, J. Using PROMETHEE to assess bioenergy pathways. *Cent. Eur. J. Oper. Res.* **2019**, *27*, 287–309. [CrossRef]

24. Myllyviita, T.; Leskinen, P.; Lähtinen, K.; Pasanen, K.; Sironen, S.; Kähkönen, T.; Sikanen, L. Sustainability assessment of wood-based bioenergy—A methodological framework and a case-study. *Biomass Bioenergy* **2013**, *59*, 293–299. [CrossRef]

25. Kontu, K.; Rinne, S.; Olkkonen, V.; Lahdelma, R.; Salminen, P. Multicriteria evaluation of heating choices for a new sustainable residential area. *Energy Build.* **2015**, *93*, 169–179. [CrossRef]

26. Alanne, K.; Salo, A.; Saari, A.; Gustafsson, S.I. Multi-criteria evaluation of residential energy supply systems. *Energy Build.* **2007**, *39*, 1218–1226. [CrossRef]

27. Campisi, D.; Gitto, S.; Morea, D. An evaluation of energy and economic efficiency in residential buildings sector: A multi-criteria analysis on an Italian case study. *Int. J. Energy Econ. Policy* **2018**, *8*, 185–196.

28. Fishburn, P.C. *Utility Theory for Decision Making*; Wiley: New York, NY, USA, 1970.

29. Hwang, C.L.; Yoon, K. *Multiple Attribute Decision Making—Methods and Applications—A State-of-the-Art Survey*; Springer: Berlin, Germany, 1981.

30. Roy, B. Classement et choix en présence de points de vue multiples (La méthode ELECTRE). *Rev. Fr. Inform. Rech. Opér.* **1968**, *2*, 57–75.

31. Brans, J.P.; Vincke, P. A preference ranking organisation method (The PROMETHEE method for multiple criteria decision-making). *Manag. Sci.* **1985**, *31*, 647–656. [CrossRef]

32. Zangheri, P.; Armani, R.; Pietrobon, M.; Pagliano, L.; Fernández Boneta, M.; Müller, A. *Heating and Cooling Energy Demand and Loads for Building Types in Different Countries of the EU*; Deliverable 2.3. of WP2 of the Entranze Project; Polytechnic University of Turin, End-Use Efficiency Research Group: Turin, Italy, 2014.

33. Ministry of Public Works, Transport and Communications. Decree-law n° 80/2006 of 4 April corresponding to the Regulation of product characteristics for thermal performance in buildings. *D. Repúb.* **2016**, *67*, 2468–2513.

34. Huijbregts, M.A.J.; Steinmann, Z.J.N.; Elshout, P.M.F.M.; Stam, G.; Verones, F.; Vieira, M.D.M.; Zijp, M.; van Zelm, R. *ReCiPe 2016: A Harmonized Life Cycle Impact Assessment Method at Midpoint and Enpoint Level—Report 1: Characterization*; National Institute for Public Health and the Environment: Bilthofen, The Netherlands, 2016.

35. Petrusa, J.; Norris, S.; Depro, B. *Regulatory Impact Analysis (RIA) for Proposed Residential Wood Heaters NSPS Revision*; U.S. Environmental Protection Agency: Washington, DC, USA, 2014.

36. Short, W.; Packey, D.J.; Holt, T. *A Manual for the Economic Evaluation of Energy Efficiency and Renewable Energy Technologies*; National Renewable Energy Laboratory: Golden, CO, USA, 1995.

37. García-Gusano, D.; Espegren, K.; Lind, A.; Kirkengen, M. The role of the discount rates in energy systems optimisation models. *Renew. Sustain. Energy Rev.* **2016**, *59*, 56–72. [CrossRef]

38. Steinbach, J.; Staniaszek, D. *Discount Rates in Energy System Analysis*; Buildings Performance Institute Europe: Berlin, Germany, 2015.

39. Association of Forestry Producers of Portugal. Forest Markets. Available online: http://www.apfc.pt/ (accessed on 18 October 2018).

40. Eurostat, Energy Prices. Available online: https://ec.europa.eu/eurostat/web/energy/data/database (accessed on 18 October 2018).

41. Chau, J.; Sowlati, T.; Sokhansanj, S.; Preto, F.; Melin, S.; Bi, X. Techno-economic analysis of wood biomass boilers for the greenhouse industry. *Appl. Energy* **2009**, *86*, 364–371. [CrossRef]

42. Dias, A.C.; Arroja, L.; Capela, I. Carbon dioxide emissions from forest operations in Portuguese eucalypt and maritime pine stands. *Scand. J. For. Res.* **2007**, *22*, 422–432. [CrossRef]

43. Hunsberger, R.; Mosey, G. *Pre-Feasibility Analysis of Pellet Manufacturing on the Former Loring Air Force Base Site*; National Renewable Energy Laboratory: Golden, CO, USA, 2014.

44. Dos Santos, A.J.R. *Work-Related Accidents in Portugal: Contributions to the Improvement of Prevention Effectiveness*; University of Algarve: Faro, Portugal, 2017.

45. Instituto da Conservação da Natureza e das Florestas (ICNF). *IFN6—Áreas dos Usos do Solo e das Espécies Florestais de Portugal Continental. Resultados Preliminare*; Instituto da Conservação da Natureza e das Florestas: Lisboa, Portugal, 2013.

46. Dias, L.C.; Domingues, A.R. On multi-criteria sustainability assessment: Spider-gram surface and dependence biases. *Appl. Energy* **2014**, *113*, 159–163. [CrossRef]

47. Domingues, A.R.; Marques, P.; Garcia, R.; Freire, F.; Dias, L.C. Applying Multi-Criteria Decision Analysis to the Life-Cycle Assessment of Vehicles. *J. Clean. Prod.* **2015**, *107*, 749–759. [CrossRef]

48. Dias, L.C.; Passeira, C.; Malça, J.; Freire, F. Integrating Life-Cycle Assessment and Multi-Criteria Decision Analysis to compare alternative biodiesel chains. *Ann. Oper. Res.* **2016**, 1–16. [CrossRef]

49. Figueira, J.; Greco, S.; Roy, B.; Slowinski, R. An overview of ELECTRE methods and their recent extensions. *J. Multi Criteria Decis. Anal.* **2013**, *20*, 61–85. [CrossRef]

50. Bouyssou, D. Ranking methods based on valued preference relations: A characterization of the net flow method. *Eur. J. Oper. Res.* **1992**, *60*, 61–67. [CrossRef]

51. Prado-Lopez, V.; Seager, T.P.; Chester, M.; Laurin, L.; Bernardo, M.; Tylock, S. Stochastic multiattribute analysis (SMAA) as an interpretation method for comparative life-cycle assessment (LCA). *Int. J. Life Cycle Assess.* **2014**, *19*, 405–416. [CrossRef]

52. Butler, J.; Jia, J.; Dyer, J. Simulation techniques for the sensitivity analysis of multi-criteria decision models. *Eur. J. Oper. Res.* **1997**, *103*, 531–546. [CrossRef]

53. United Nations. *Transforming Our World: The 2030 Agenda for Sustainable Development*; United Nations: New York, NY, USA, 2015.

Article

A Systematic Approach to Predict the Economic and Environmental Effects of the Cost-Optimal Energy Renovation of a Historic Building District on the District Heating System

Vlatko Milić [1,*], Shahnaz Amiri [1,2] and Bahram Moshfegh [1,2]

[1]   Division of Energy Systems, Department of Management and Engineering, Linköping University,
      581 83 Linköping, Sweden; shahnaz.amiri@liu.se (S.A.); bahram.moshfegh@liu.se (B.M.)
[2]   Division of Building, Energy and Environment Technology, Department of Technology and Environment,
      University of Gävle, 801 76 Gävle, Sweden
*   Correspondence: vlatko.milic@liu.se; Tel.: +46-1328-4751

Received: 2 December 2019; Accepted: 3 January 2020; Published: 6 January 2020

Abstract: The economic and environmental performance of a district heating (DH) system is to a great extent affected by the size and dynamic behavior of the DH load. By implementing energy efficiency measures (EEMs) to increase a building's thermal performance and by performing cost-optimal energy renovation, the operation of the DH system will be altered. This study presents a systematic approach consisting of building categorization, life cycle cost (LCC) optimization, building energy simulation and energy system optimization procedures, investigating the profitability and environmental performance of cost-optimal energy renovation of a historic building district on the DH system. The results show that the proposed approach can successfully be used to predict the economic and environmental effects of cost-optimal energy renovation of a building district on the local DH system. The results revealed that the financial gains of the district are between 186 MSEK (23%) and 218 MSEK (27%) and the financial losses for the DH system vary between 117–194 MSEK (5–8%). However, the suggested renovation measures decrease the local and global CO_2 emissions by 71–75 metric ton of $CO_{2eq.}$/year (4%) and 3545–3727 metric ton of $CO_{2eq.}$/year (41–43%), respectively. Total primary energy use was decreased from 57.2 GWh/year to 52.0–52.2 GWh/year.

Keywords: LCC optimization; building energy simulation; energy system optimization; energy renovation; historic building district; district heating system

1. Introduction

Fossil fuel supply sources dominate the European building heat market, representing approximately 66% of the total end-use heat demand [1]. The total final energy use in the residential and services sector in Sweden in 2017 was 146 TWh, according to the Swedish Energy Agency [2]. Electricity and oil represent 50% and 8% of the final energy use in the residential and services sector, respectively. Substituting oil and electricity as sources of energy for heating systems with efficient use of resources via district heating (DH) is, therefore, vital in order to achieve a sustainable energy system in the building sector. DH is a heat distribution system where heat is produced at a central plant and distributed via the DH system to end-users. It is common to cogenerate the production of heat with electricity production, i.e., combined heat and power plant (CHP). Benefits from DH include the possibility to use different fuels, using waste that would otherwise be sent to landfill, cogeneration with electricity production, energy security and high supply security.

The profitability and environmental performance of a DH system are directly connected to the buildings' energy use within the DH system. In Sweden, DH represents 32% of the final energy use in the residential and services sector [2], and there is a significant potential to increase its share. However, future heat loads in DH systems are complex to predict, due to, among other aspects, the degree of energy renovations in the building stock. Strong incentives exist for building owners to perform building energy renovation in the form of economic savings and environmental benefits [3]. From an energy savings perspective, it is especially important to study the historic building stock because of the generally poorer thermal performance of older buildings compared to newer ones [4]. An example of investigating the energy savings potential in historic building districts includes the work presented in Liu et al. [5] using the historic district in Visby, Sweden, as a case study. The district was connected to the municipality's DH system. A combination of building categorization and life cycle cost (LCC optimization) was used. The results showed a possible decrease of 31% in energy use and LCC when targeting LCC optimum. It is important to note that no investigation was performed for the effects on the surrounding DH system from the suggested renovation measures. By implementing energy efficiency measures (EEMs) in the buildings, the heat demand will be reduced in the DH system, which is counterproductive for the DH supplier. On the other hand, EEMs could also be a beneficial measure for the DH supplier by reducing the utilization of peak load plants during wintertime, with a high operation cost on these days [6]. As a result, the economic and environmental influences on a DH system from performing building energy renovation are complex to predict because of varying local conditions in terms of fuel mix, CHP plant, heat-only production boiler etc. In addition, to overcome difficulties during studies of complex energy systems, such as cities, there is a need for efficient and rational use of computational software [7].

There are a number of scientific investigations addressing the impacts on the local energy systems from building energy renovation. Åberg and Widén [8] investigated the impact of implementing assumed EEMs in residential buildings in six different DH systems in Sweden. This was performed using a cost-optimization model structure. It was stated that a decrease in heat demand, due to energy efficiency in residential buildings results in reduced use of fossil fuels and biomass in the DH system. Moreover, it was found that the decrease and reduction of heat demand, as a result of the implementation of EEMs, mainly affect heat-only production boiler. In fact, in five out of six DH systems, the quantity of CHP-generated electricity per unit of produced heat is improved. The same cost-optimization tool was used during an investigation of the entire Swedish DH sector based on four pre-defined DH systems [9]. The four DH systems were used to describe a DH sector in aggregated form. The objective was to investigate the effects of reductions in heat demand, due to building energy efficiency improvement. It was concluded that heat demand reductions, for the most part, decrease global CO_2 emissions and the use of biomass and fossil fuels. However, to maximize the reduction in CO_2 emissions, the heat production technologies in different DH systems should be taken into account. Lundström and Wallin [10] also highlighted that by decreasing heat demand through the insulation of the building envelope, the heat load curve is levelled out, resulting in decreased greenhouse gas emissions and improved energy efficiency. The study object consisted of two multi-family buildings in Eskilstuna, Sweden, from the 1960s and 1970s. Le Truong et al. [11] investigated the effects of heat- and electricity-saving measures in multistory concrete-framed and wood-framed versions of an existing residential building connected to DH in Växjö, Sweden. The measures included domestic hot water reduction, improved building thermal envelope, ventilation heat recovery and higher household appliance efficiency. Energy savings from these measures were calculated using building energy simulation (BES) software. It was concluded that measures that decrease more peak load production also give higher primary energy savings. The largest primary energy savings were obtained from efficient household appliances. The importance of decreasing electricity use to reduce primary energy use is in line with the findings from Lidberg et al. [12], which are based on systematic studies of the energy renovation of a multi-family house connected to the DH system. Environmental benefits in the form of decreased global CO_2 emissions from electricity savings were also found by Difs et al. [13], together with economic benefits for the local energy system. The investigation was performed using an energy system optimization model where the energy conservation measures were implemented one at a time, with the town of Linköping, Sweden, as a case study. Åberg and Henning [6] also studied the DH network in Linköping using an energy system optimization model, with a focus on impacts from energy savings in existing residential buildings built during the period 1961–1980. It was concluded that reductions in heat demand in the studied building stock result in decreased global use of fossil fuels and global CO_2 emissions, which is in accordance with the results of Difs et al. [13] (based on a similar model of the DH system in Linköping). It was also shown that it is primarily heat-only production that decreases when the heat demand is reduced, which supports the results from Åberg and Widén [8]. A similar study was performed by Lidberg et al. [14] using the city of Borlänge, also situated in Sweden, as a case study where four energy efficiency packages were investigated. The results showed that electricity production decreases, due to building energy renovation, with less electricity imported to the market as a result. In addition, it was concluded that global greenhouse gas emissions are decreased for all packages, because of the assumption that biomass is a possible replacement for fossil fuels elsewhere.

As presented above, there are a number of scientific investigations addressing the impacts on the DH system from building energy renovation. However, research on the effects of the cost-optimal energy renovation of a building district with regard to the consequential impact on the local energy systems is scarce. The objective of this study is to present a systematic approach with a systematic perspective when investigating the impact of cost-optimal energy renovation of a historic building district concerning economics and environmental performance in terms of primary energy use and CO_2 emissions on the DH system. A novel combination of building categorization, LCC optimization, -BES and energy system optimization procedures is the foundation for the proposed research. The approach is applicable for aggregating LCC and heating load to clusters of buildings and districts. Hence, it is possible to reflect the dynamic behavior of individual buildings, clusters and building districts before and after cost-optimal renovation, and the consequential effect on the surrounding DH system. Consequently, the contribution to the research community consists of the development of an effective and useful approach for predicting economic and environmental effects of the optimal renovation of buildings, clusters and districts connected to the DH system. Moreover, the present study will provide a systematic and holistic overview of the connections between building energy performance, profitability and environmental impact in terms of the CO_2 emissions of a DH system located in a Northern European climate during cost-optimal building energy renovation.

2. Systems Approach and Computational Tools

In this study, a systematic approach is used to predict the effects of cost-optimal energy renovation of a building district on the DH system. Firstly, representative building types are obtained through categorization of a building district which is the historic district in Visby, Sweden, in the current research. The original energy use and LCCs of the building types are calculated using the LCC optimization software OPtimal Energy Retrofit Advisory-Mixed Integer Linear Program (OPERA-MILP). By using OPERA-MILP, the cost-optimal energy renovation strategy is also obtained for each building type. The renovation strategy includes cost-efficient EEMs, for example, insulation of the building envelope and window replacement, as well as airtightness. BES software IDA ICE is then used to model and simulate each building type in order to obtain the building heat demand over time and heat load duration curve for each building type, before and after energy renovation. The heat load for the various building types, as well as energy use and LCCs, can be aggregated at cluster level and district level. The heat load for the DH system is thereafter converted into a flexible time division suitable for larger DH systems by using the software Converter [15]. Lastly, based on the converted heat load, the effects of the energy renovations performed on the DH system in the form of environmental impact (CO_2 emissions and primary energy use), optimal DH production and system cost are calculated using the energy systems optimization model MODEST (Model for Optimisation of Dynamic Energy Systems with Time-dependent components and boundary conditions). The proposed approach is illustrated in Figure 1.

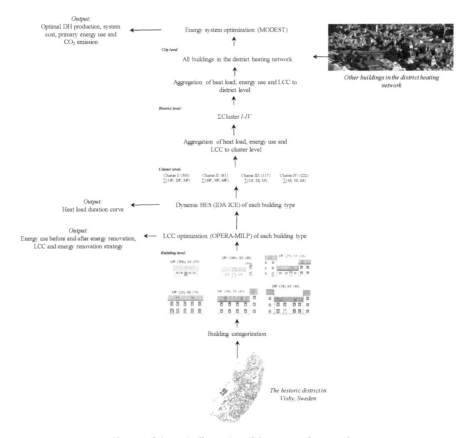

Figure 1. Schematic illustration of the proposed approach.

2.1. Life Cycle Cost Optimization in Buildings: OPERA-MILP

The in-house LCC optimization software OPERA-MILP is used to obtain the cost-optimal energy renovation package for the various building types. OPERA-MILP has been used successfully in a number of previous scientific investigations, e.g., [16–19]. A specified period of time is set for the optimization in the OPERA-MILP software, which is 50 years in this study. Costs related to investments in the heating system, EEMs targeting the building envelope, energy costs and maintenance costs for building components are all taken into consideration. The total LCC of a building is calculated according to Equation (1).

$$\text{LCC}_{\text{building}} = \text{LCC}_{\text{investment}} + \text{LCC}_{\text{energy}} + \text{LCC}_{\text{maintenance}} - \text{RV}, \tag{1}$$

where $\text{LCC}_{\text{building}}$ = total building LCC over the optimization period, $\text{LCC}_{\text{investment}}$ = total investment costs for EEMs targeting the building envelope and heating system, $\text{LCC}_{\text{energy}}$ = energy cost over the specified period of time, $\text{LCC}_{\text{maintenance}}$ = maintenance cost for building components and RV = residual value of the investment costs connected to EEMs on the building envelope, heating system and maintenance performed on the building.

The implemented EEMs targeting the building envelope in OPERA-MILP include replacing windows, weatherstripping, floor insulation, roof insulation and inside and outside insulation of the external walls. Concerning heating systems, DH, groundwater heat pump, electric radiators and wood boiler are incorporated into the software. The costs for the various measures in OPERA-MILP are calculated based on cost functions, see Equations (2)–(5). The use of cost functions for describing the costs for the various measures allows for calculating a mathematical optimum, and hence, optimization of LCC.

$$C_{ws.} = C_1 \cdot m, \tag{2}$$

$$C_{w.} = C_2 \cdot A_{window}, \tag{3}$$

$$C_{i.m.} = C_3 \cdot A_{b.c} + C_4 \cdot A_{b.c} + C_5 \cdot A_{b.c} \cdot t, \tag{4}$$

$$C_{h.s.} = C_6 + C_7 \cdot P_{h.s.} + C_8 \cdot P_{h.s.}, \tag{5}$$

The cost for weatherstripping is dependent on the number of windows, see Equation (2) where $C_{ws.}$ = total cost for weatherstripping, C_1 = the weatherstripping cost per window and m = number of windows in the building. Meanwhile, the cost for replacing windows is dependent on the window area, see Equation (3) where $C_{w.}$ = total cost for window replacement, C_2 = window replacement cost per m^2 and A_{window} = total window area. Equation (4) presents the cost function for the insulation measures where $C_{i.m.}$ = total cost for the insulation measure, C_3 = maintenance or inevitable cost per m^2, $A_{b.c}$ = total area of the building component, C_4 = fixed part of the insulation cost per m^2, C_5 = variable insulation cost per m^2 depending on insulation thickness and t = insulation thickness. Equation (5) shows the cost function for the installation of a heating system. $C_{h.s.}$ = total installation cost for the heating system, C_6 = base cost for the heating system not depending on power, C_7 = cost depending on the power of the heating system, $P_{h.s}$ = maximum power of the heating system and C_8 = cost for piping system depending on the power of the heating system.

The building's energy balance is calculated based on a time resolution of 12 time steps where each step corresponds to a month during a year. The energy balance of the building includes heat losses in the form of transmission, ventilation and infiltration and hot water use, as well as heat gains in the form of solar gains and heat from internal sources including electrical appliances, building occupants and heat from processes, such as cooking. A utilization factor for the internal heat gains energy is also considered. The maximum heat power demand is calculated based on the preset indoor temperature, the outdoor design temperature of Visby for the specific building type and the total heat losses of the building. In addition, the power demand for domestic hot water is taken into account.

2.2. Building Energy Simulation: IDA ICE

IDA ICE is a commercial software program within the field of BES. The mathematical models are written in Neutral Model Format (NMF) code, enabling the user to make changes in the models. The software allows a dynamic whole-year simulation. The energy balance in IDA ICE is calculated depending on building geometry, solar radiation, internal heat loads, HVAC (heating, ventilation and air-conditioning) conditions and building construction data.

2.3. Energy System Optimization: MODEST

In this paper, an optimization model is known as MODEST [9,20,21] is used to model and analyze the DH system in Visby, and to investigate the effects of the performed energy renovations of a historic building district on the DH system. MODEST has a flexible time division, which can reflect demand peaks and diurnal, weekly and seasonal variations in energy demand and other parameters, e.g., fuel and electricity prices. MODEST has been applied to electricity and DH systems for approximately 50 local utilities [22–24]. The model has been used in numerous scientific investigations, e.g., [6,13,14,25–29]. A thorough description of MODEST is given by Henning [29] and Henning et al. [24].

With the use of MODEST it is possible to calculate the net income of the DH system, see Equation (6).

$$\text{Net income}_{DH\ system} = DH\ income - System\ cost, \tag{6}$$

where Net income$_{DH\ system}$ = net income over the optimization period for the DH system, DH income = income for sold DH to end-users and System cost = total cost for the optimal DH production. It should be noted that only costs connected to the optimal DH production are considered, and not other expenditures for running the DH system such as employee salaries.

3. Description of the Historic District and the District Heating System in Visby

3.1. The Historic District

Visby is a town located in southeastern Sweden on the island of Gotland, about 100 km east of the mainland in the Baltic Sea, with approximately 24,000 inhabitants. The average annual outdoor temperature in Visby is +7.7 °C. Twelve historic residential building types, which are typical historic buildings in Visby, are selected as the study object [30,31]. The building types are obtained based on a categorization study of the historic district of Visby. The categorization method can be divided into three main steps:

1. Inventory of the building stock, i.e., gathering and compilation of building data;
2. Categorization (allocating buildings in groups depending on the number of adjoining walls, number of stories and floor area);
3. Selection of building types that are representative of the building stock (each building type selected based on average values of various building characteristics).

The categorization method resulted in a total of 12 building types: 1W–6W and 1S–6S ("W" indicating a building structure of wood and "S" indicating a building structure of stone). Building types 1W–3W and 1S–3S represent single-family houses with one story and a heated attic floor, and building types 4W–6W and 4S–6S multi-family buildings with two stories and a heated attic floor. Moreover, other differences between the building types include building thermal envelope performance, basement type, adjoining walls, etc. The building types are illustrated in Figure 2 where a photograph of the corresponding building category is also shown below each illustration. Building category 1 is seen in the top left corner, building category 2 in the top center and so forth.

Using the 12 building types described above, Liu et al. [5] formed four clusters based on variations in building size and type of building structure. Single-family houses 1W–3W formed Cluster *I*, multi-family buildings 4W–6W Cluster *II*, single-family houses 1S–3S Cluster *III* and multi-family buildings 4S–6S Cluster *IV*. Cluster *I* includes 500 similar single-story wood buildings, Cluster *II* 81 similar multi-story wood buildings, Cluster *III* 117 similar single-story stone buildings and Cluster *IV* 222 similar multi-story stone buildings. Construction data for the building types is given in Table 1, as well as the number of buildings in each cluster. In all building types, the majority of the window area faces east and west from the building, with double-glazed windows. All building types are naturally ventilated. The indoor temperature is set to 21 °C following the recommendations by the Public Health Agency of Sweden [32]. Internal heat generation and domestic hot water use are estimated using data from Sveby, a development program for companies and organizations in the construction and real estate industry. The use of domestic hot water is differentiated whether the building is a single-family house (Cluster *I* and Cluster *III*) or a multi-family building (Cluster *II* and Cluster *IV*).

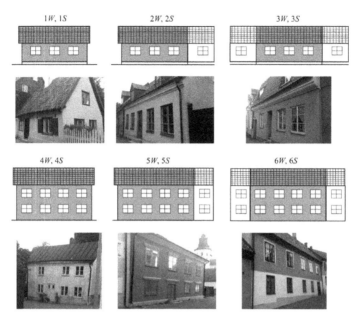

Figure 2. The studied building types with photographs of corresponding building categories.

Table 1. The number of buildings in each cluster and construction data for the building types.

Cluster		I						II			III			IV
Building Type		1W	2W	3W	4W	5W	6W	1S	2S	3S	4S	5S	6S
No. of Buildings		309	166	25	33	30	18	55	46	16	75	83	64
Building structure	Wood	×	×	×	×	×	×						
	Stone							×	×	×	×	×	×
Basement type	Crawl space	×	×	×				×	×	×			
	Unheated basement				×	×	×				×	×	×
No. of adjoining walls		0	1	2	0	1	2	0	1	2	0	1	2
External walls	Area (m²)	86	61	45	245	180	116	80	57	43	235	173	112
	U-value (W/(m²·°C))	0.65	0.65	0.65	0.67	0.67	0.67	1.8	1.8	1.8	1.97	1.97	1.97
Windows	Area (m²)	12	12	12	44	37	30	12	12	12	44	37	30
	U-value (W/(m²·°C))	2.9	2.9	2.9	2.9	2.9	2.9	2.9	2.9	2.9	2.9	2.9	2.9
Roof	Area (m²)	71	79	92	170	159	159	65	73	86	161	150	150
	U-value (W/(m²·°C))	0.18	0.18	0.18	0.25	0.25	0.25	0.18	0.18	0.18	0.25	0.25	0.25
Floor	Area (m²)	49	50	58	133	124	129	44	44	52	123	115	120
	U-value (W/(m²·°C))	1.10	1.10	1.10	0.23	0.23	0.23	1.10	1.10	1.10	0.23	0.23	0.23
Heated area (m²)		98	100	116	398	372	387	87	88	104	369	345	360
Heated volume (m³)		216	219	256	942	881	917	192	194	228	874	817	852
Air change rate (ACH)		0.76	0.74	0.72	0.65	0.64	0.62	0.77	0.75	0.73	0.65	0.64	0.62

To investigate the impact from cost-optimal energy renovation of a historic building district on the DH system in Visby, three different cases concerning LCC and building energy use are investigated. DH is set as the default heating system in all cases. To enable an assessment of the effects of energy renovation, a reference case for the studied buildings is modelled. The remaining lifetime of the building components is set to 0 years in all cases. In Case 1 (the reference case), no EEMs on the building envelope are allowed. DH is set as the default heating system since it is the most common heating form in Sweden and is available in Visby. In Case 2 (LCC optimum), the lowest LCC is obtained by selecting cost-effective EEMs on the building envelope. In Case 3, specific energy targets are achieved for the studied building types (83 kWh/m^2 and 79 kWh/m^2 for the single-family houses and multi-family buildings, respectively) according to Swedish building regulations, BBR. It should be noted that the energy targets vary depending on geographical location and heating system type in BBR. The location of Visby and DH as the preset heating system are, therefore, considered for the energy targets in Case 3. The cases included in this investigation are summarized in Table 2.

Table 2. Summary of the investigated cases in this study. LCC, life cycle cost.

LCC/Energy Target	EEMs on the Building Envelope	Case No.
Reference	Not allowed	Case 1
LCC optimum	Allowed	Case 2
Swedish building regulations—83 kWh/m^2 and 79 kWh/m^2 for single-family houses and multi-family buildings, respectively	Allowed	Case 3

3.2. The District Heating System

Heat generation is carried out by energy utilities which belong to Gotlands Energi AB (GEAB), the municipal energy utility for Visby. GEAB provides approximately 185 GWh/year (normal year corrected using Energy-Index from the Swedish Meteorological and Hydrological Institute (SMHI) heat to approximately 1250 end-users through the DH distribution network. The end-users of heat, i.e., the customers, can be small single-family houses, large multi-family buildings or various types of public buildings, such as libraries and schools. The total length of the DH pipe network is 90 km, and the culvert heat losses are approximately 11%. The supply temperature varies between 75 and 100 °C depending on the outdoor temperature. A schematic description of the DH system in Visby connected to end users, including Cluster *I*, Cluster *II*, Cluster *III* and Cluster *IV*, with heat production facilities is shown in Figure 3. The DH production is dominated by biomass. Most of the DH production takes place in heat-only biomass boilers (HOB 5, HOB 6) with flue gas condensation (FGC) together with a compressor heat pump (HP). There are also a number of heat-only peak load boilers, namely bio oil boilers (HOB 1, HOB 2, HOB 3, HOB 4), an electric boiler (HOB 8) and an oil boiler (HOB 7), which are only in operation during the winter season. On average, the heat-only bio-fuel boilers (HOB 5, HOB 6) produce about 90% of DH demand in the system. In addition, landfill gas is utilized to produce heat in the DH system.

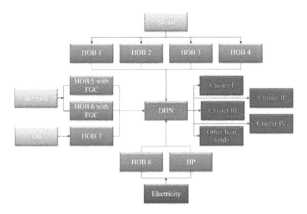

Figure 3. Schematic of the DH system in Visby.

4. Input Data

4.1. Input Data: OPERA-MILP

The modelling in terms of building properties was based on the data presented in Section 3.1 for each building. The LCC optimization of the various building types is performed based on a time period of 50 years. The remaining lifetime of the building components is set to zero, resulting in an inevitable cost occurring for the various building elements. DH is set as the heating system before and after renovation. The remaining lifetime of the DH units in the building types is set to zero.

The selection of a cost-optimal energy renovation strategy is directly dependent on the input data used in the OPERA-MILP software. Costs for the various EEMs incorporated into OPERA-MILP are analyzed using cost functions in OPERA-MILP, see Section 2.1. The cost functions are developed based on the Swedish database Wikells [33], which provides up-to-date market costs, as well as using manufacturer data. The investment costs for the various EEMs are given in Table 3. Since the twelve building types include buildings with a structure of either wood or stone, investment costs are developed for both building structures. The minimum insulation thickness is set to 2 cm and the maximum to 42 cm with a step resolution of 2 cm. The thermal conductivity of additional insulation is 0.037 W/(m·°C). In addition, the cost for weatherstripping varies depending on whether the building is a single-family house or a multi-family building. This is because of a difference in window size. The estimated U-values for the windows are 1.5 W/(m²·°C), 1.2 W/(m²·°C) and 0.8 W/(m²·°C) for the double-glazed, triple-glazed and triple-glazed + low emission windows, respectively. It should be noted that window replacement is inevitable in Case 2 and Case 3. However, since no EEMs are allowed in the reference case, see Table 2, a maintenance cost for the windows is included corresponding to the investment cost for the double-glazed windows. The lifetime is set at 50 years for all insulation measures and 30 years for windows [34]. The lifetime for weatherstripping is assumed to be 10 years.

Table 3. Investment cost for the energy efficiency measures (EEMs).

EEMs	$C_{1,\,SFH}$[1]/MFB[2] (SEK/Window)	$C_{2,\,DG}$[3]/TG[4]/TG+LE[5] (SEK/m² Window)	$C_{3,\,wood/stone}$ (SEK/m²)	$C_{4,\,wood/stone}$ (SEK/m²)	$C_{5,\,wood/stone}$ (SEK/m²·m)	C_6 (SEK)	C_7 (SEK/kW)	C_8 (SEK/kW)
Weatherstripping	441/617	-	-	-	-	-	-	-
Window replacement	-	6738/8492/12,169	-	-	-	-	-	-
Roof insulation	-	-	0/0	0/0	679/679	-	-	-
Floor insulation	-	-	0/0	242/242	799/799	-	-	-
External wall inside insulation	-	-	153/153	908/1335	1267/1267	-	-	-
External wall outside insulation	-	-	407/407	2411/2571	1267/1267	-	-	-
DH unit	-	-	-	-	-	22,611	415	255

[1] SFH = single-family house, [2] MFB = multi-family building, [3] DG = double-glazed, [4] TG = triple-glazed and [5] TG + LE = triple-glazed + low emission glass.

The exchange rate is set to 10.30 SEK ≈ 1 Euro [35]. A discount rate of 5% is used [36]. For piping system in the heating systems, a lifetime of 50 years is set. Data concerning fuel prices, annual cost, life times and efficiencies connected to the DH unit is presented in Table 4. Fuel prices and annual costs for DH are obtained from Gotlands Energi AB using data from 2016.

Table 4. Price data, life time and efficiency for the DH unit.

Heating System Data	Fuel Price (SEK/MWh)	Annual Cost (SEK)	η (-)	Life Time (Years)
DH unit	959	315	0.95	25 [37]

4.2. Input Data: IDA ICE

In the present study, version 4.8 of IDA ICE was used. The buildings were modeled using climate data based on ASHRAE IWEC2 [38]. Each building was modelled based on the data presented in

Section 3.1. The simulations were performed during one year with 14 days of dynamic startup in order to achieve stability in the thermal characteristics of the building, such as the set indoor temperature, which is also the default in the software. IDA ICE models visualizing each building category are seen in Figure 4. Building category 1 is seen in the top left corner, building category 2 in the top center, and so forth.

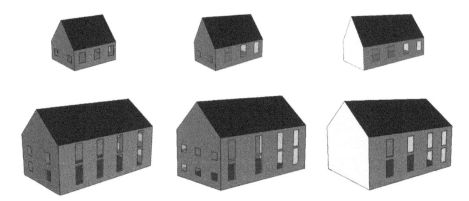

Figure 4. The buildings modeled in IDA ICE software.

4.3. Input Data: MODEST

The DH system in Visby is modeled in MODEST. Technical data for utilities, CO_2 emission factors and flexible time divisions that can reflect peaks and diurnal, weekly and seasonal variations in DH demand in Visby are given as input data for MODEST, see Table 5. The DH demand (185 GWh/year) varies continuously throughout weekdays, weekends and months. Table 6 shows the MODEST time periods used in this study. CO_2 emission factors for the fuel used in the DH model include both production and transportation. The electricity prices in the model reflect the average value of the actual Swedish electricity prices on the Nord Pool spot market during 2018 for Visby including electricity distribution costs and electricity tax. A period of 50 years is studied, and the discount rate is set to 5%. The primary energy factor for the DH produced in Visby is set as 0.31 since this is the local value [39].

Table 5. DH production plants and their properties.

Heat-Only Boilers/Heat Pump	Heat Production (MW)	Fuel	CO$_2$ Emission Factor [40,41] (g CO$_{2eq}$./kWh)
HOB 1	27.2	Bio oil	5
HOB 2	10.8	Bio oil	5
HOB 3	6	Bio oil	5
HOB 4	11.8	Bio oil	5
HOB 5 with FGC [1]	10	Biomass	11
HOB 6 with FGC [1]	18	Biomass	11
HOB 7	6.6	Oil	290
HOB 8 [2]	17	Electricity	969
Heat pump (HP) [3]	12	Electricity	969

[1] $\eta = 1.10$, [2] $\eta = 0.98$, [3] COP = 2.5.

Table 6. The MODEST-time periods applied in this study.

Month	Days and Hours	Month	Days and Hours
November–March	Mon.–Fri., 6–7	April–October	Mon.–Fri., 6–22
	Mon.–Fri., 7–8		Mon.–Fri., 22–6
	Mon.–Fri., 8–16		Sat., Sun. and holiday, 6–22
	Mon.–Fri., 16–22		Sat., Sun. and holiday, 22–6
	Mon.–Fri., 22–6		
	Sat., Sun. and holidays, 6–22		
	Sat., Sun. and holidays, 22–6		
	Top day, 6–7		
	Top day, 7–8		
	Top day, 8–16		
	Top day, 16–22		
	Top day, 22–6		

The DH production is dominated by biomass. Most of the DH production takes place in heat-only biomass boilers (HOB 5, HOB 6) with flue gas condensation (FGC) together with a compressor heat pump (HP). There are also a number of heat-only peak load boilers, namely bio oil boilers (HOB 1, HOB 2, HOB 3, HOB 4), an electric boiler (HOB 8) and an oil boiler (HOB 7), which are only in operation during the winter season. On average, the heat-only bio-fuel boilers (HOB 5, HOB 6) produce about 90% of DH demand in the system.

The marginal electricity production accounting model has been used in order to calculate global CO_2 emissions. This means that a coal-fired condensing power plant has been assumed to be the short-term marginal power plant in the European electricity system. According to marginal electricity, the production of 1 GWh electricity gives 969 metric ton of $CO_{2eq.}$ [40]. Hence, the local electricity used, e.g., for heat pumps, will increase the electricity produced by coal-fired condensed power plants and that the global CO_2 emissions will, therefore, increase.

5. Results and Discussion

5.1. Energy Use, LCC, and System Cost, Net Income and Environmental Effects of the DH System before Energy Renovation of the Studied Buildings in Visby

The following section presents energy use and LCC for the buildings before energy renovation. This is shown at building type level, cluster level and district level. In addition, the environmental effects and system cost are given at city level together with the total net income for the DH system.

5.1.1. Building Level

The original performance of the building types in terms of specific energy use and LCC has been predicted using OPERA-MILP. Energy use and LCC for the various building types are presented in Table 7. The specific energy varies between 99.1 and 200.1 kWh/m^2 for the wood buildings and between 143.2 and 324.0 kWh/m^2 for the stone buildings. The overall better thermal performance of the wood buildings compared to the stone buildings, as a result of the lower U-value of the external walls as presented in Table 1, is the reason for the lower energy use in the wood buildings. It should be noted that the specific energy use (heating and domestic hot water use) for buildings built before 1940 in Sweden is on average 125 kWh/m^2 for single-family houses and 146 kWh/m^2 for multi-family buildings [42]. This means that all single-family houses in this study have higher energy use (in the range of 161.2–324.0 kWh/m^2) compared to the national average. The opposite trend is seen with the multi-family buildings, where the buildings in Cluster *II*, i.e., building types 4W–6W, and building type 6S have a lower specific energy use (varying between 99.1 and 143.2 kWh/m^2) compared to the Swedish average.

Table 7. Maximum building power demand, specific energy use and LCC for the various building types.

Cluster	*I*			*II*			*III*			*IV*		
Building Type	1W	2W	3W	4W	5W	6W	1S	2S	3S	4S	5S	6S
Maximum power demand (kW)	6.9	6.4	6.7	19.0	16.3	14.4	9.1	7.8	7.6	27.3	22.2	17.9
Specific energy use (kWh/m²)	200.1	178.6	161.2	128.1	115.4	99.1	324.0	266.2	218.0	219.8	187.3	143.2
Specific LCC (kSEK/m²)	5.6	5.0	4.4	3.7	3.3	2.7	8.1	6.8	5.6	5.5	4.8	3.6

In terms of specific LCC during the optimization period of 50 years, the LCC is in the range between 2.7 and 5.6 kSEK/m² (kSEK stands for thousands of SEK) for the wood buildings and between 3.6 and 8.1 kSEK/m² for the stone buildings. There is a strong correlation between high/low energy use and high/low LCC. The reason for this is that the LCC before energy renovation consists only of energy cost and heating system installation cost, where the energy cost constitutes the largest expenditure of LCC by a significant degree because of the low installation cost for the building's heating system, i.e., the DH system, see Table 3.

5.1.2. Cluster Level

Energy use and LCC for the four building clusters are presented in Table 8. The specific energy use for the various clusters is 190.7 kWh/m², 117.1 kWh/m², 284.9 kWh/m² and 185.8 kWh/m² for Cluster *I*, Cluster *II*, Cluster *III* and Cluster *IV*, respectively. The corresponding figures are 207.1 kWh/m² for all single-family houses (Cluster *I* and Cluster *III*) and 166.4 kWh/m² for all multi-family buildings (Cluster *II* and Cluster *IV*), which is 82 kWh/m² and 20 kWh/m² above the national average for single-family houses and multi-family buildings, respectively.

Table 8. Energy use and LCC for the four building clusters before renovation.

Cluster	*I*			*II*			*III*			*IV*		
Building Type	1W	2W	3W	4W	5W	6W	1S	2S	3S	4S	5S	6S
No. of Buildings	309	166	25	33	30	18	55	46	16	75	83	64
Specific energy use (kWh/m²)	200.1	178.6	161.2	128.1	115.4	99.1	324.0	266.2	218.0	219.8	187.3	143.2
Specific energy use at the cluster level (kWh/m²)	190.7			117.1			284.9			185.8		
Total energy use at the cluster level (GWh)	9.5			3.7			3.0			14.7		
Specific LCC (kSEK/m²)	5.6	5.0	4.4	3.7	3.3	2.7	8.1	6.8	5.6	5.5	4.8	3.6
Specific LCC at Cluster level (kSEK/m²)	5.3			3.3			7.2			4.7		
Total LCC at Cluster level (MSEK)	263.8			103.2			75.6			372.9		

The specific LCCs during the optimization period of 50 years are 5.3 kSEK/m², 3.3 kSEK/m², 7.2 kSEK/m² and 4.7 kSEK/m² for Cluster *I*, Cluster *II*, Cluster *III* and Cluster *IV*, respectively. Hence, Cluster *II* has the lowest specific LCC (the cluster with the lowest specific energy use) and Cluster *III* the highest specific LCC, which is also the cluster with the highest specific energy use. Cluster *II* and Cluster *IV* have moderate specific LCCs. The two clusters also have moderate specific energy use. When comparing the total LCCs of the various clusters, Cluster *I* and Cluster *IV* have total LCCs of 264 MSEK (MSEK stands for millions of SEK) and 373 MSEK, which is significantly higher compared to Cluster *II* (103 MSEK) and Cluster *III* (76 MSEK). This is largely explained by the large heated areas in these two clusters (Cluster *I* ~49,800 m² and Cluster *IV* ~79,400 m²) compared to Cluster *II* and Cluster *III*, which have heated areas of ~31,300 m² and ~10,500 m², respectively.

Heating load duration curves for the various clusters are constructed using hourly data obtained through the energy simulations of the building types in IDA ICE, see Figure 5. The duration curve visualizes the heating load in descending order in terms of magnitude, considering the periods of time during which the loads occur. Therefore, the duration curve directly reflects the thermal performance of the clusters, as well as the heated area in each cluster. Consequently, Cluster *IV* has the highest heat load followed by Cluster *I*, Cluster *II* and Cluster *III*. Furthermore, the duration curves also provide information about the baseload that is visualized by the lowest loads in the diagram. For Cluster *I* to Cluster *IV*, the baseload occurs approximately between 1600 h and 2600 h.

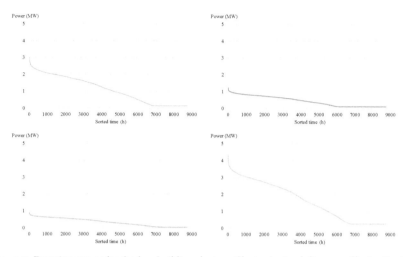

Figure 5. Duration curves for the four building clusters. Cluster *I* = top left corner, Cluster *II* = top right corner, Cluster *III* = bottom left corner and Cluster *IV* = top right corner.

5.1.3. District Level

The total heated area before renovation for the 920 buildings in the studied district, i.e., the four building clusters, is 0.17 km². The energy use at the district level is calculated at 31 GWh. The corresponding figure in terms of specific energy use for the district is 180.8 kWh/m². The total LCC is 816 MSEK during an analysis period of 50 years, and the specific LCC is 4.8 kSEK/m². The power demand over the year for the studied district and the corresponding load duration curve are shown in Figure 6. The peak load for the district is 9.4 MW, and the baseload is 0.45 MW. It is important to note that due to the monthly time-step calculation procedure in OPERA-MILP, average monthly internal heat gains are also used in IDA ICE for comparability purposes. However, a study by Milić et al. [43] showed that the predictions of energy usage correspond to a maximum annual difference of 4% when considering varying internal heat gains. The low impact from varying internal heat gains is explained by that the case study consisted of buildings with overall poor thermal performance and low time constant, similar to the buildings in the present research.

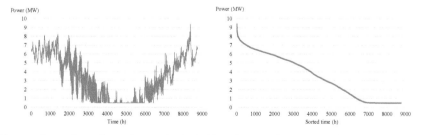

Figure 6. Power demand over the year for the building district with a corresponding duration curve.

5.1.4. City Level

To enable a comparison to be made before and after cost-optimal energy renovation of the 920 buildings in the study object, the performance of the DH network in Visby before building renovation is presented in the following section. As mentioned in Section 3.2, the total DH demand before renovation is 184.6 GWh/year for Visby. The primary energy use is 57.2 GWh/year considering the local primary energy factor for Visby (0.31). The peak load for the city is 54 MW.

The optimal DH production by the various plants in Visby is shown in Figure 7 for Case 1 (before renovation of Clusters *I–IV*) using the optimization model MODEST.

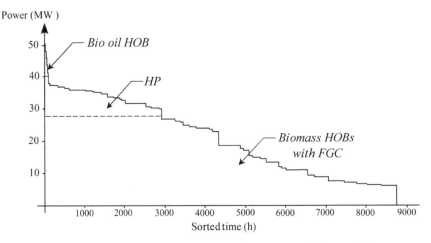

Figure 7. Optimal DH production in Visby before renovation of Clusters *I–IV*.

In Case 1 (i.e., the Reference case), no EEMs targeting the building envelope are introduced in the building stock and the DH demand is produced by the utilities heat-only boilers and heat pumps, see Figure 7. Biomass heat-only boilers (HOB 5, HOB 6) generate the largest part of the DH demand (164.7 GWh/year). Heat pumps additional supply heat of 18.0 GWh/year to the DH system. The bio oil boilers produce the rest of the DH demand (0.5 GWh/year) during the peak load. In addition, landfill gas supplies around 1.4 GWh/year to the DH system. Local and global $CO_{2eq.}$ emissions, in this case, are 1667 metric ton/year and 8648 metric ton/year, respectively. The system cost for the DH system is 39.8 MSEK/year and 727 MSEK during an optimization period of 50 years. The system cost is the present value of capital costs, fixed costs, costs related to the output power and costs associated with the amount of energy used (energy costs). The revenue from sold DH to end-users is 3232 MSEK resulting in a net income of 2505 MSEK for the DH system, or 137 MSEK/year.

5.2. Cost-Optimal Energy Renovation; EEMs, Energy Use, LCC and System Cost, Net Income and Environmental Effects of the DH System

This section presents the results from cost-optimal energy renovation in terms of selected EEMs, energy use and LCC for the studied cases in this investigation. This is given at building type level, cluster level and district level. Furthermore, an assessment of the effects of cost-optimal renovation on the Visby DH system in terms of system cost, net income, primary energy use and CO_2 emissions is presented.

5.2.1. Building Level

Using LCC optimization, cost-optimal energy renovation strategies are identified for LCC optimum (Case 2) and energy targets of 83 kWh/m^2 and 79 kWh/m^2 for single-family houses and multi-family buildings, respectively, according to Swedish building energy regulations (Case 3). The selection of EEMs targeting the building envelope for the various cases is presented in Table 9. The effects of energy renovation in terms of specific energy use and LCC are given in Table 10. It is important to note that replacing windows is inevitable in Case 2 and Case 3 because the remaining lifetime of the building elements is zero, resulting in weatherstripping as a side effect, since the new windows are assumed to be airtight. From a profitability point of view, double-glazed windows are, in most cases, the suggested window type.

Table 9. Selected EEMs targeting the building envelope.

Cluster		I			II			III			IV		
Building Type		1W	2W	3W	4W	5W	6W	1S	2S	3S	4S	5S	6S
Window type	Case 2	DG*	DG	DG	DG	DG	DG	DG	DG	DG	DG	DG	DG
	Case 3	DG	DG	DG	DG	DG	DG	DG	DG	DG	DG	DG	DG
Floor insulation	Case 2	26	26	26	0	0	0	24	24	24	0	0	0
	Case 3	24	32	26	0	0	0	24	24	24	0	0	0
Roof insulation	Case 2	12	12	12	18	16	16	10	10	10	16	16	16
	Case 3	10	18	4	24	24	6	0	0	0	16	16	10
External wall inside insulation	Case 2	0	0	0	0	0	0	20	20	20	20	20	20
	Case 3	8	2	0	6	4	0	20	14	6	12	8	4

* DG = double-glazed window.

Table 10. Specific energy use and LCC for the various building type. The percentage change in Case 2 and Case 3, compared to Case 1 is indicated in parentheses with an italic font.

Cluster		I			II			III			IV		
Building Type		1W	2W	3W	4W	5W	6W	1S	2S	3S	4S	5S	6S
Case 1	kWh/m²	200.1	178.6	161.2	128.1	115.4	99.1	324.0	266.2	218.0	219.8	187.3	143.2
	kSEK/m²	5.6	5.0	4.4	3.7	3.3	2.7	8.1	6.8	5.6	5.5	4.8	3.6
Case 2	kWh/m²	111.5 *(−44)*	93.5 *(−48)*	80.2 *(−50)*	97.6 *(−24)*	88.0 *(−24)*	76.4 *(−23)*	79.3 *(−78)*	72.3 *(−74)*	67.8 *(−70)*	73.5 *(−68)*	69.1 *(−65)*	64.7 *(−56)*
	kSEK/m²	4.3 *(−24)*	3.7 *(−26)*	3.1 *(−28)*	3.2 *(−14)*	2.9 *(−14)*	2.4 *(−14)*	5.6 *(−37)*	4.7 *(−35)*	3.8 *(−34)*	4.0 *(−32)*	3.5 *(−30)*	2.7 *(−27)*
Case 3	kWh/m²	81.8 *(−61)*	81.4 *(−55)*	83.1 *(−48)*	77.6 *(−41)*	75.9 *(−36)*	78.7 *(−21)*	83.1 *(−77)*	82.0 *(−71)*	82.1 *(−63)*	77.6 *(−66)*	76.9 *(−60)*	77.5 *(−46)*
	kSEK/m²	4.7 *(−20)*	4.1 *(−18)*	3.2 *(−28)*	3.5 *(−8)*	3.2 *(−7)*	2.4 *(−13)*	5.7 *(−36)*	4.8 *(−34)*	3.9 *(−32)*	4.0 *(−31)*	3.5 *(−28)*	2.9 *(−21)*

It is important to be aware that the cost-optimal energy renovation strategy is unique for each building type because of unique building conditions in the form of layout and construction. In any case, the strategies in terms of selected insulation measures are very similar for the building types in each cluster as the building properties are highly similar. The selection of 26 cm floor insulation and 12 cm roof insulation in Case 2 for all building types in Cluster *I*, i.e., single-family houses in wood, is an example of this. Other trends that can be seen in Table 10 concerning selected EEMs in the building types and clusters are:

- Floor insulation in the range between 24 cm and 32 cm is profitable for Cases 2 and 3 in Cluster *I* and Cluster *III*, i.e., building types standing on crawl space, because of high transmission losses originally

- Roof insulation is generally profitable in all clusters and cases because of low retrofit costs (despite an originally low U-value). The suggested insulation thickness varies between 10 and 18 cm at LCC optimum (Case 2). The corresponding figure for the energy target according to the Swedish building regulations (Case 3) for the building types varies more, due to the cost-effective comparison between EEMs on the building envelope
- Inside insulation of the external walls is profitable for all cases in the stone buildings, Cluster *III* and Cluster *IV*, because of a high U-value before renovation, 1.80–1.97 W/(m²·°C). The suggested insulation thickness is 20 cm in Case 2, but varies between 2 and 20 cm in Case 3, due to the cost-effective comparison between EEMs. The inside insulation of the external walls is also necessary in some of the wooden buildings to achieve the energy targets in Case 3.

The energy use at the cost-optimum point, Case 2, varies between 80.2–111.5 kWh/m², 76.4–97.6 kWh/m², 67.8–79.3 kWh/m² and 64.7–73.5 kWh/m² for Cluster *I*, Cluster *II*, Cluster *III* and Cluster *IV*, respectively. The percentage decrease in energy use is the highest for the building types standing on crawl space (Cluster *I* and Cluster *III*) and the building types with an external wall of stone (Cluster *III* and Cluster *IV*). The reason for this is the additional insulation of these building elements, as well as the poor U-value before renovation. Of the single-family houses, building type 3W and all building types in Cluster *III*, 1S–3S, achieve the Swedish building regulations target of 83 kWh/m² at the cost-optimum point. Concerning the multi-family buildings at LCC optimum, building type 6W, and all building types in Cluster *IV*, 4S–6S, achieve the energy target of 79 kWh/m². Hence, the specific energy use at LCC optimum is lower than the energy target in the Swedish building regulation for all building types in stone, i.e., Cluster *II* and Cluster *IV*. In most optimizations, the energy target in Case 3 is achieved in Case 2, only requiring cost-effective comparison between the EEMs on the building envelope. This is, however, not the case in building types 1W, 2W, 4W and 5W where the energy use is further decreased in Case 3 to reach the energy target of 83 kWh/m² according to the Swedish building regulations (BBR). Concerning LCC, the costs during an optimization period of 50 years are lowered by 14–37% at LCC optimum compared to before renovation. In Case 3, the LCC is lowered for all buildings compared to before renovation varying between 8% and 36%. The largest percentage decrease in LCC occurs for the building types where the energy use has been decreased the most. For instance, the LCC is decreased the most in the building types in Cluster *III* (34–37% at LCC optimum) which are also the building types with the highest percentage decrease in energy use (70–78% at LCC optimum). The same tendencies are identified for the buildings with the lowest percentage decrease in energy use, 23–24% for the building types in Cluster *II*. The decreases in LCC are determined at 14% for the building types in Cluster *II*.

5.2.2. Cluster Level

Specific energy use and LCC in Cases 1–3 for the four building clusters are presented in Table 11, with the percentage difference after renovation (Cases 2 and 3) compared to Case 1 given in the parenthesis. The specific energy use for the various clusters varies between 69.3 kWh/m² and 103.7 kWh/m² in Case 2. The corresponding figures for Case 3 are 77.2 and 82.5 kWh/m². The specific LCC varies between 2.9–5.0 kSEK/m² and 3.1–5.1 kSEK/m² for Case 2 and Case 3, respectively. Of the study's building clusters and cases, Cluster *III* is the cluster with the highest percentage decrease in energy use (76%) and LCC (31%) compared to before renovation at the cost-optimum point. Cluster *II* has the lowest percentage decrease in energy use (23%), as well as in LCC (12%). The low and high percentage decreases in energy use for Cluster *II* and Cluster *III* are explained by the originally good thermal performance of the building types in Cluster *II* and the poor thermal performance of the building types in Cluster *III*. Furthermore, it is shown that energy renovation according to the energy target in Case 3 does not result in a higher LCC compared to before renovation in any of the optimizations. In fact, the LCC is decreased by 0.2–2.1 kSEK/m², or 6–29%.

Table 11. Specific energy use and LCC in Cases 1–3 for the various clusters.

Cluster	I			II			III			IV		
Building Type	1W	2W	3W	4W	5W	6W	1S	2S	3S	4S	5S	6S
No. of Buildings	309	166	25	33	30	18	55	46	16	75	83	64
Energy use before renovation—Case 1 (kWh/m²)	190.7			117.1			284.9			185.8		
LCC optimum—Case 2 (kWh/m²)	103.7 (−46%)			89.4 (−23%)			74.7 (−76%)			69.3 (−64%)		
Swedish building regulations—Case 3 (kWh/m²)	81.7 (−57%)			77.2 (−35%)			82.5 (−73%)			77.3 (−59%)		
LCC before renovation—Case 1 (kSEK/m²)	5.3			3.3			7.2			4.7		
LCC optimum—Case 2 (kSEK/m²)	4.0 (−24%)			2.9 (−12%)			5.0 (−31%)			3.4 (−28%)		
Swedish building regulations—Case 3 (kSEK/m²)	4.4 (−17%)			3.1 (−6%)			5.1 (−29%)			3.5 (−26%)		

Annual energy use and total LCC during the optimization period of 50 years for the four building clusters are shown in Figure 8 for Cases 1–3. Cluster *IV* has the highest energy use and LCC in all three cases, followed by Cluster *I*, Cluster *II* and Cluster *III*. The annual energy use is 4.0–9.5 GWh, 2.4–3.7 GWh, 0.7–3.0 GWh and 5.3–14.8 GWh for Cluster *I*, Cluster *II*, Cluster *III* and Cluster *IV*, respectively. The corresponding figure for LCC is 200–265 MSEK, 91–104 MSEK, 49–76 MSEK and 260–373 MSEK. When investigating the profitability of the suggested renovation measures between the clusters, there is a clear difference in terms of potential decreases in total LCC. Cluster *I* and Cluster *IV* show significant financial gains at the cost-optimum point (Case 2) compared to before energy renovation. The LCC is decreased from 265 MSEK to 200 MSEK in Cluster *I* and from 373 MSEK to 260 MSEK in Cluster *IV*. It is important to note the corresponding decrease in total energy use from cost-optimal energy renovation of Clusters *I* and *IV*, which is calculated at 13.8 GWh. This is 10.6 GWh more compared to the total decrease in Cluster *II* and Cluster *III* (total energy use = 6.7 GWh in Case 1 and 3.5 GWh in Case 2). The consequential effect from the suggested renovation measures with a decreased heating load on the local DH system will be addressed in Section 5.2.4. With regard to the energy targets in Case 3 (Swedish building regulations), the energy use is decreased from 1.3 GWh (Cluster *II*) to 8.7 GWh (Cluster *IV*). Following the tendency in terms of decrease in energy use, Cluster *IV* has the highest LCC savings in Case 3 (101 MSEK). Clusters *I–III* also show financial gains when performing renovation according to the energy targets in Case 3, varying between 8 and 51 MSEK.

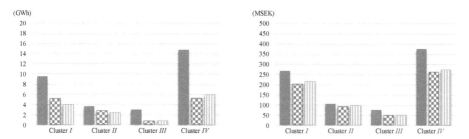

Figure 8. Annual energy use (to the left) and total LCC (to the right) during an optimization period of 50 years for Clusters *I–IV* in Case 1 (solid blue), Case 2 (red checkered) and Case 3 (green vertical lines).

5.2.3. District Level

The following section presents the effects at district level from the cost-optimal renovation of the various building clusters in terms of LCC and energy use. Consequently, the findings can be used as guidance for various stakeholders when investigating the possibilities for energy renovating building districts. The total energy use and LCC for the 920 buildings in the studied district, i.e., the four building clusters, are shown in Table 12. Total energy use is decreased in the range between 17.0 GWh and 17.7 GWh compared to before renovation. The highest decrease occurs when performing energy renovation according to the Swedish building regulations (Case 3), where the energy use is decreased by 57%. Interestingly, the percentage difference in energy use in Cases 2, 3 compared to Case 1 varies between 55% and 57%, showing that the energy use after renovation only varies slightly between Case 2 and Case 3. In terms of expenditure during the 50 years, LCC is calculated at 818 MSEK, 600 MSEK and 632 MSEK for Cases 1–3, respectively. Hence, financial revenue between 186 and 218 MSEK are made possible through renovation according to the cases in this study.

Table 12. Energy use and LCC of the building district in Cases 1–3.

Case	Energy Use (GWh)	LCC (MSEK)
Energy use before renovation—Case 1	30.9	818
LCC optimum—Case 2	13.9 (−55%)	600 (−27%)
Swedish building regulations—Case 3	13.2 (−57%)	632 (−23%)

The effects of the cost-optimal renovation of each cluster on the district are unique, in terms of both energy use and LCC. This is visualized in Figure 9 for energy renovation of each cluster at the cost-optimum (Case 2), arranged from the cluster with the highest energy savings, Cluster *IV*, to the cluster with the smallest energy savings (Cluster *II*) compared to before renovation. As shown in Figure 9, the LCC for the district is decreased by 113 MSEK, 64 MSEK, 28 MSEK and 13 MSEK when renovating Cluster *IV*, Cluster *I*, Cluster *III* and Cluster *II*, respectively. The corresponding annual energy savings are 9.5 GWh, 4.3 GWh, 2.3 GWh and 0.9 GWh. This shows clear differences between the clusters in terms of cost-optimal energy efficiency potential and a potential decrease in LCC. Furthermore, using the proposed approach, it is possible to rank building clusters in a district based on the profitability of suggested energy renovation measures, or energy savings. Hence, various stakeholders can apply and use the approach to obtain the optimal outcome for building energy renovation depending on both financial budget and final objectives.

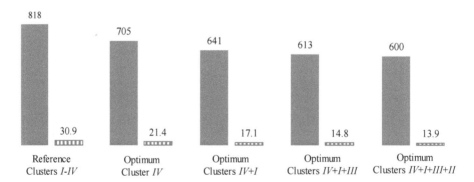

Figure 9. LCC MSEK (solid blue) and total energy use GWh (red vertical lines) before renovation and when selecting cost-optimal EEMs in the various clusters at LCC optimum (Case 2).

The corresponding effect, i.e., renovation of the various clusters according to the description above for Figure 9, on the DH load for the historic district can be seen in a duration curve in Figure 10 using the time steps in MODEST. The top duration curve shows the duration curve for Case 1, and the second, third, fourth and fifth curves from the top show the duration curve after cost-optimal renovation (Case 2) of Cluster *IV*, Clusters *IV* + *I*, Clusters *IV* + *I* + *III* and Clusters *IV* + *I* + *III* + *II*, respectively.

Following the tendencies in terms of decreases in energy use and LCC, renovating Cluster *IV* and Cluster *I* corresponds to the largest decreases in heat load in the duration curve compared to Case 1. This can be exemplified by analyzing the peak load for each duration curve. The peak load (9.4 MW) is decreased by 2.6 MW, 1.3 MW, 0.6 MW and 0.3 MW when energy renovating Cluster *IV*, Cluster *I*, Cluster *III* and Cluster *II*, respectively.

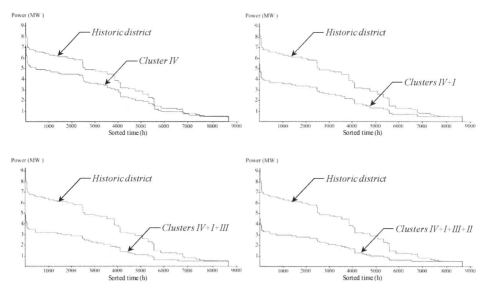

Figure 10. Duration curve when selecting cost-optimal EEMs in the various cluster at LCC optimum (Case 2), as well as before renovation (Case 1) for the historic district. Top left = duration curve for the district in Case 1 and after renovating Cluster *IV*, top right = duration curve after renovating Clusters *IV* + *I*, bottom left = duration curve after renovating Clusters *IV* + *I* + *III* and bottom right = duration curve after renovating Clusters *IV* + *I* + *III* + *II*, respectively.

5.2.4. City Level

This section presents the optimization results for the Visby DH system after cost-optimal energy renovation of the building stock. In addition, the effects on the Visby DH system from the cost-optimal renovation of the various building clusters (when arranged by decreased energy use compared to before renovation) have also been investigated and the optimization results are described in this section. In addition, the LCC and environmental impacts from the renovation of a historic district are shown. Table 13 summarizes the results for Cases 1–3. Table 14 summarizes the results when carrying out energy renovation separately for each building cluster at the cost-optimum point (Case 2), ranging from the cluster with the highest energy savings to the cluster with the lowest energy savings, i.e., corresponding to the following order: Cluster *IV*, Clusters *IV* + *I*, Clusters *IV* + *I* + *III* and Clusters *IV* + *I* + *III* + *II*. The results consist of the DH production, biomass/bio oil supply and electricity used in DH system, local and global CO_2 emissions, system cost and net income for the Visby DH system.

Table 13. DH production, biomass/bio oil supply and electricity used in the DH system, local and global CO_2 emissions, system cost and net income for the Visby DH system for all cases.

DH Production, CO_2 Emissions, System Cost and Net Income	Case 1	Case 2	Case 3
DH (GWh/year)	184.6	168.4	167.6
Energy supply (GWh/year)			
Biomass	149.7	143.4	143.0
Electricity	7.2	3.6	3.4
Bio oil	0.6	0.3	0.3
System cost (MSEK/year)	39.8	34.9	29.9
System cost over 50 years (MSEK)	726.7	637.3	546.0
Net income DH system (MSEK)	2505	2311	2388

Table 14. DH production, biomass/bio oil supply and electricity used in the DH system, local and global CO_2 emissions, system cost and net income for the Visby DH system after cost-optimal energy renovation of the various building clusters.

DH Production, CO_2 Emissions, System Cost and Net Income	Case 1	Cluster IV	Clusters IV + I	Clusters IV + I + III	Clusters IV + I + III + II
DH (GWh/year)	184.6	175.7	171.4	169.3	168.4
Energy supply (GWh/year)					
Biomass	149.7	146.4	144.6	144	143.4
Electricity	7.2	5.1	4.2	3.8	3.6
Bio oil	0.6	0.5	0.4	0.4	0.33
Local $CO_{2\,eq.}$ emissions (ton/year)	1667	1630	1610	1600	1596
Global $CO_{2\,eq.}$ emissions (ton/year)	8648	6606	5718	5285	5203
System cost (MSEK)	726.7	675.5	653.6	640.8	637.3
Net income DH system (MSEK)	2505	2400	2347	2323	2311

Cost-optimal energy renovation when targeting LCC optimum (Case 2) and Swedish building regulations (Case 3) affects the performance of the DH system in Visby. As shown in Table 13, biomass heat-only boilers generate the largest part of the DH in all the studied cases. Heat pump and bio oil boilers, as well as landfill gas, supply additional heat during the year in the all studied cases. The largest overall change in terms of environmental performance and system cost of the DH system compared to before energy renovation occurs in Case 3, where DH production by the utilities biomass, bio oil heat-only boilers and heat pumps decrease the most compared with all other cases. The electricity used in the DH system, in this case, is 3.4 GWh/year. The corresponding figures for biomass and bio oil supply are 143.0 GWh and 0.3 GWh, respectively. The system cost is reduced the most in Case 3, calculated at 9.9 MSEK/year because of the reduced fuel supply and the use of electricity, as shown in Table 13. The system cost in Case 2 is 34.9 MSEK/year compared to 39.8 MSEK/year before energy renovation. When considering the net income of the DH system, it can be seen that the net income is decreased from 2505 MSEK (Case 1) to between 2311 MSEK (Case 2) and 2388 MSEK (Case 3), due to less DH sold to end-users.

The result of cost-optimal energy renovation when targeting LCC optimum (Case 2) in the various clusters compared with Case 1 shows that selecting cost-optimal EEMs in Clusters IV + I + III + II decreases the system cost the most compared to the other alternatives presented in Table 14, i.e., separately energy renovating each cluster ranging from the cluster with the highest energy savings to the cluster with the lowest energy savings. The system cost is decreased by 51.2 MSEK, 73.1 MSEK, 85.9 MSEK and 89.4 MSEK over a period of 50 years when renovating Cluster IV, Clusters IV + I, Clusters IV + I + III and Clusters IV + I + III + II, respectively. However, as stated in the section above, a result of renovating the various clusters less DH is sold to end-users, and the revenue is decreased. The net income of the DH system is 2505 MSEK in Case 1, and 2400 MSEK, 2347 MSEK, 2323 MSEK and 2311 MSEK when energy renovating Cluster IV, Clusters IV + I, Clusters IV + I + III and Clusters IV + I + III + II, respectively. The local and global CO_2 emissions are decreased by 37–71 metric ton and 2042–3445 metric ton, respectively, as presented in Table 14.

By calculating the revenue for the sold DH and the system cost, as well as LCC and energy use for the various clusters, it is possible to predict the overall outcome of cost-optimal renovation of a historic district in Visby in terms of economics and environmental impact. The LCC for the district and the net income for the DH system over the course of 50 years, as well as annual figures for primary energy use and local and global CO_2 emissions, are presented in Figure 11. The net income is 2388 MSEK for the DH system when renovating according to Swedish energy targets (Case 3) compared to the 2311 MSEK at the cost-optimum point (Case 2), which is due to the decrease in electricity use. It is important to mention that the revenue for the DH system is decreased for both Case 2 and Case 3 compared to Case 1 by 194 MSEK and 117 MSEK, respectively. This is because less DH is sold to the end-users. The sold DH to end-users generates an income of 3232 MSEK, 2948 MSEK and 2934 MSEK over a period of 50 years in Case 1, Case 2 and Case 3, respectively. However, there are significant environmental benefits from the suggested renovations. Regarding local and global CO_2 emissions, Case 3 corresponds to the lowest emissions of all investigated cases as a result of the largest decrease in used and supplied fuel in the DH system. Local and global $CO_{2eq.}$ emissions are calculated at 1592 metric ton/year (4% decrease compared to Case 1) and 4921 metric ton/year (43% decrease compared to Case 1), respectively. The primary annual energy use in the city decreases by 5.0 GWh and 5.2 GWh for Case 2 and Case 3, respectively, compared to before renovation (57.2 GWh), which corresponds to a 9% decrease. A reduction in CO_2 emissions, due to decreased heat load is supported by earlier research findings, e.g., [9,10].

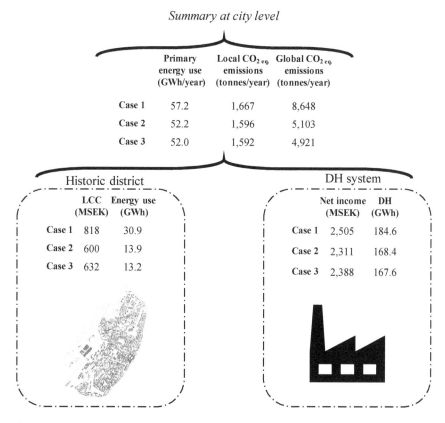

Figure 11. Summary of the LCC for the district and the net income for the DH system over the course of 50 years, as well as annual primary energy use and local and global CO_2 emissions.

There are several interesting aspects to study in future work. When analyzing the effects on the local DH system from the energy renovation of a historic district, it is important to be aware of the time-dependent behavior of the energy systems. This includes larger populations living in cities, resulting in higher resource utilization [44], and expected higher outdoor temperature. A larger population in Visby will most likely increase the heat load of the DH system, due to more buildings being connected to the local energy systems. However, as an effect of global warming, higher outdoor temperatures will occur, resulting in a lower heat load in the DH system. As shown by Shen et al. [45], global climate change also affects the optimal energy renovation strategy for a building and varies depending on location. Other factors that can influence the operation of the DH system include changes in fuel prices, for both the DH and competing for heating supply systems, as well as technology development and CO_2 on fossil fuels [46]. The aforementioned factors are all of interest to study in future work. Concerning the replicability of the results to other cities and energy systems, the authors would like to point out that it is difficult to generalize the effects of cost-optimal energy renovation. This is due to the fact that each building district is unique in terms of dynamic behaviour, and the operation strategy, as well as the fuel mix of the DH system. Hence, the authors recommend performing separate case studies on various local DH systems in cold climate to quantify economic and environmental indicators using the proposed approach. Moreover, another aspect of interest to study is the economic and environmental effects of the expansion to a CHP system. This is especially the case since there is an ongoing process of electrifying the buildings, transport and industry sectors.

6. Conclusions

The majority of the research investigations in the field of building LCC optimization have tended to focus on the building owners' perspective. This includes financial gains and improving buildings' environmental performance as a result of cost-optimal energy renovation. However, a connection exists between the energy renovation of DH heated buildings and the performance of the DH system. There is currently a knowledge gap regarding the effects on the local DH system in terms of profitability and environmental performance from cost-optimal energy renovation of a building district. This research proposes a systematic approach to study the effects on DH system of the above-mentioned parameters from cost-optimal energy renovation of a building district. The study object consists of 920 buildings located in Visby, Sweden, divided into twelve building types and four building clusters. Three different cases are included in the study: Case 1 = no EEMs on the building envelope corresponding to the reference case, Case 2 = LCC optimum (the cost-optimum point) and Case 3 = energy renovation according to Swedish building regulations, BBR.

The results of this study show that by using the proposed approach, it is possible to predict the effects of cost-optimal building renovation concerning economics and environmental performance on the local DH system. In terms of economics, it is revealed that suggested energy renovations in the historic district correspond to financial gains by 186–218 MSEK (23–27%). However, cost-optimal renovation of the historic district decreases the revenue of the DH system by 117–194 MSEK (5–8%), due to less DH sold to end-users. There is, on the other hand, a clear connection between building renovation and decreased local and global CO_2 emissions, due to decreased use and supply of fuel, which confirms the findings of previous research within the field. This is especially true when carrying out energy renovation according to BBR, resulting in local and global decreases of 75 metric ton of $CO_{2eq.}$/year (4%) and 3727 metric ton of $CO_{2eq.}$/year (43%), respectively. Concerning the effects of building renovation of the various clusters in the district, it is seen that the total heated area of each cluster is highly significant, as well as the current thermal performance. Cost-optimal renovation with no preset energy target (Case 2) corresponds to a decrease in energy use of 13.8 GWh (57%) and financial gains of 178 MSEK (28%) for Cluster *I* and Cluster *IV*, where the two clusters are characterized by overall low thermal performance and represent 76% of the total heated area in the studied district. Meanwhile, the corresponding figures for Cluster *II* and Cluster *III* are 3.2 GWh (47%) and 40 MSEK (22%). Lastly, the suggested renovation strategies are unique to each building type, due to differences in layout and construction. However, the suggested measures for the buildings in each cluster are rather similar because building properties are comparable to a high degree. This can be exemplified through the selection of 24–36 cm floor insulation in Cluster *I* and Cluster *III* because of a high *U*-value and the selection of inside insulation of the external walls with 20 cm in Cluster *I* and Cluster *III* at the cost-optimum point, which is also explained by the initially poor thermal performance of the external walls.

Author Contributions: B.M. was the project leader. S.A. was responsible for the optimizations performed in energy system optimization software MODEST and helped with writing the paper connected to the MODEST results. V.M. performed and analyzed the optimization and simulation procedures in OPERA-MILP and IDA ICE, and wrote the paper. B.M. contributed with valuable advice and revision of the manuscript. All authors have read and agreed to the published version of the manuscript.

Funding: This research was funded by of the Swedish Energy Agency, grant number P31669-3 and P44335-1.

Acknowledgments: The authors thank Carolin Boelin and Anna Ekman from GEAB for providing useful data about the DH system in Visby. Furthermore, we would like to thank the Swedish Energy Agency for providing financial support [grant number P31669-3 and P44335-1].

Conflicts of Interest: The authors declare no conflict of interest.

References

1. U. Persson and S. Werner. Available online: http://www.heatroadmap.eu/resources/STRATEGO% 20WP2%20-%20Background%20Report%204%20-%20Heat%20%26%20Cold%20Demands.pdf (accessed on 9 November 2017).
2. The Swedish Energy Agency. *Energy in Sweden 2019. An Overview*; The Swedish Energy Agency: Stockholm, Sweden, 2019; ISSN 1404-3343.
3. Artola, I.; Rademaekers, K.; Williams, R.; Yearwood, J. *Boosting Building Renovation: What potential and value for Europe? 2016, PE 587.326*; European Union: Brussels, Belgium, 2016.
4. National Board of Housing, Building and Planning. *Energy Use in Buildings-Technical Characteristics and Calculations-Results from the Project BETSI*; National Board of Housing, Building and Planning: Karlskrona, Sweden, 2010; ISBN 978-91-86559-83-0.
5. Liu, L.; Rohdin, P.; Moshfegh, B. Investigating cost-optimal refurbishment strategies for the medieval district of Visby in Sweden. *Energy Build.* **2017**, *158*, 750–760. [CrossRef]
6. Åberg, M.; Henning, D. Optimisation of a Swedish district heating system with reduced heat demand due to energy efficiency measures in residential buildings. *Energy Policy* **2011**, *39*, 7839–7852. [CrossRef]
7. Manfren, M.; Caputo, P.; Costa, G. Paradigm shift in urban energy systems through distributed generation: Methods and models. *Appl. Energy* **2011**, *88*, 1032–1048. [CrossRef]
8. Åberg, M.; Widén, J. Development, validation and application of a fixed district heating model structure that requires small amounts of input data. *Energy Convers. Manag.* **2013**, *75*, 74–85. [CrossRef]
9. Åberg, M. Investigating the impact of heat demand reductions on Swedish district heating production using a set of typical system models. *Appl. Energy* **2014**, *118*, 246–257. [CrossRef]
10. Lundström, L.; Wallin, F. Heat demand profiles of energy conservation measures in buildingsand their impact on a district heating system. *Appl. Energy* **2016**, *161*, 290–299. [CrossRef]
11. Le Truong, N.; Dodoo, A.; Gustavsson, L. Effects of heat and electricity saving measures in district-heated multistory residential buildings. *Appl. Energy* **2014**, *118*, 57–67. [CrossRef]
12. Lidberg, T.; Gustafsson, M.; Myhren, J.A.; Olofsson, T.; Ödlund, L. Environmental impact of energy refurbishment of buildings within different district heating systems. *Appl. Energy* **2018**, *227*, 231–238. [CrossRef]
13. Difs, K.; Bennstam, M.; Trygg, L.; Nordenstam, L. Energy conservation measures in buildings heated by district heating-A local energy system perspective. *Energy* **2010**, *35*, 3194–3203. [CrossRef]
14. Lidberg, T.; Oloffson, T.; Trygg, L. System impact of energy efficient building refurbishment within a district heated region. *Energy* **2016**, *106*, 45–53. [CrossRef]
15. Division of Energy Systems at Linköping University. *Converter*; Linköping University: Linköping, Sweden, 2016.

16. Liu, L.; Rohdin, P.; Moshfegh, B. LCC assessments and environmental impacts on the energy renovation of a multi-family building from the 1890s. *Energy Build.* **2016**, *133*, 823–833. [CrossRef]
17. Broström, T.; Eriksson, P.; Liu, L.; Rohdin, P.; Ståhl, F.; Moshfegh, B. A method to assess the potential for and consequences of energy retrofits in Swedish historic buildings. *Hist. Environ. Policy Pract.* **2014**, *5*, 150–166. [CrossRef]
18. Gustafsson, S.-I. Mixed integer linear programming and building retrofits. *Energy Build.* **1998**, *28*, 191–196. [CrossRef]
19. Gustafsson, S.-I. Optimal fenestration retrofits by use of MILP programming technique. *Energy Build.* **2001**, *33*, 843–851. [CrossRef]
20. Henning, D. Optimisation of Local and National Energy Systems. Development and Use of the MODEST Model. Doctoral Dissertation, No. 559. Division of Energy Systems, Linköping University, Linköping, Sweden, 1999.
21. Gebremedhin, A. A Regional and Industrial Co-Operation in District Heating Systems. Doctoral Dissertation, No. 849. Division of Energy Systems, Linköping University, Linköping, Sweden, 2003.
22. Sundberg, G.; Henning, D. Investments in combined heat and power plants: Influence of fuel price on cost minimised operation. *Energy Convers. Manag.* **2002**, *43*, 639–650. [CrossRef]
23. Sjödin, J.; Henning, D. Calculating the marginal costs of a district-heating utility. *Appl. Energy* **2004**, *78*, 1–18. [CrossRef]
24. Henning, D.; Amiri, S.; Holmgren, K. Modelling and optimization of electricity, steam and district heating production for a local Swedish utility. *Eur. J. Oper. Res.* **2006**, *175*, 1224–1247. [CrossRef]
25. Rolfsman, B. CO_2 emission consequences of energy measures in buildings. *Build. Environ.* **2002**, *37*, 1421–1430. [CrossRef]
26. Åberg, M.; Widén, J.; Henning, D. Sensitivity of district heating system operation to heat demand reductions and electricity price variations: A Swedish example. *Energy* **2012**, *41*, 525–540. [CrossRef]
27. Weinberger, G.; Amiri, S.; Moshfegh, B. On the benefit of integration of a district heating system with industrial excess heat: An economic and environmental analysis. *Appl. Energy* **2017**, *191*, 454–468. [CrossRef]
28. Amiri, S.; Henning, D.; Karlsson, B.G. Simulation and introduction of a CHP plant in a Swedish biogas system. *Renew. Energy* **2013**, *49*, 242–249. [CrossRef]
29. Amiri, S.; Weinberger, G. Increased cogeneration of renewable electricity through energy cooperation in a Swedish district heating system-A case study. *Renew. Energy* **2018**, *116*, 866–877. [CrossRef]
30. Berg, F. *Categorising a Historic Building Stock—An Interdisciplinary Approach*; UU-259149; Uppsala University: Uppsala, Sweden, 2015.
31. Broström, T.; Donarelli, A.; Berg, F. For the categorisation of historic buildings to determine energy saving. *Agathon Int. J. Archit. Art Des.* **2017**, *1*, 135–142.
32. Public Health Agency of Sweden. *Common Advice about Indoor Tempeatures*; FoHMFS 2014:17; Public Health Agency of Sweden: Stockholm, Sweden, 2014.
33. Wikells. Section Facts 17/18-Techno-economic compilation.
34. Adalberth, K.; Wahlström, Å. *Energy Audit of Buildings-Apartment Buildings and Facilities*; Swedish Standards Institute: Stockholm, Sweden, 2009; ISBN 978-91-7162-755-1.
35. European Central Bank. Euro Exchange Rates. Available online: https://www.ecb.europa.eu/stats/exchange/eurofxref/html/eurofxref-graph-sek.en.html (accessed on 3 September 2018).
36. Swedish Energy Agency and National Board of Housing, Building and Planning. *Proposals for National Stragegy for Energy Efficiency Renovation of Building*; Swedish Energy Agency and National Board of Housing, Building and Planning: Karlskrona, Sweden, 2013; ISBN 978-91-7563-049-6.
37. Mälarenergi. Buy a New Heat Exchanger. Available online: https://www.malarenergi.se/fjarrvarme/for-husagare/kop-ny-fjarrvarmecentral (accessed on 27 November 2018).
38. ASHRAE. ASHRAE IWEC2 Weather Files for International Locations. Available online: http://ashrae.whiteboxtechnologies.com/ (accessed on 2 March 2017).
39. Swedenergy. Environmental Assessment of District Heating. Available online: https://www.energiforetagen.se/statistik/fjarrvarmestatistik/miljovardering-av-fjarrvarme/miljovarden-fran-tidigare-ar/ (accessed on 27 November 2018).
40. Grönkvist, S.; Sjödin, J.; Westermark, M. Models for assessing net CO_2 emissions applied on on district heating technologies. *Int. J. Energy Res.* **2003**, *27*, 601–613. [CrossRef]

41. Swedenergy. *Agreement in the Heating Market Committe 2018*; Swedenergy: Stockholm, Sweden, 2018.
42. The Swedish Energy Agency. Energy Statistics. Available online: http://www.energimyndigheten.se/statistik/den-officiella-statistiken/alla-statistikprodukter/ (accessed on 20 February 2019).
43. Milić, V.; Ekelöw, K.; Moshfegh, B. On the performance of LCC optimization software OPERA-MILP by comparison with building energy simulation software IDA ICE. *Build. Environ.* **2018**, *128*, 305–319. [CrossRef]
44. Vassileva, I.; Campillo, J.; Schwede, S. Technology assessment of the two most relevant aspects for improving urban energy efficiency identified in six mid-sized European cities from case studies in Sweden. *Appl. Energy* **2017**, *194*, 808–818. [CrossRef]
45. Shen, P.; Braham, W.; Yi, Y. The feasibility and importance of considering climate change impacts in building retrofit analysis. *Appl. Energy* **2019**, *233*, 254–270. [CrossRef]
46. Le Truong, N.; Gustafsson, L. Minimum-cost district heat production systems of different sizes under different environmental and social cost scenarios. *Appl. Energy* **2014**, *136*, 881–893. [CrossRef]

Article

Estimation of Carbon Dioxide Emissions from a Diesel Engine Powered by Lignocellulose Derived Fuel for Better Management of Fuel Production

Karol Tucki [1,*], Olga Orynycz [2,*], Andrzej Wasiak [2], Antoni Świć [3], Remigiusz Mruk [1] and Katarzyna Botwińska [1]

[1]   Department of Production Engineering, Institute of Mechanical Engineering,
      Warsaw University of Life Sciences, Nowoursynowska Street 164, 02-787 Warsaw, Poland;
      remigiusz_mruk@sggw.pl (R.M.); katarzyna_botwinska@wp.pl (K.B.)
[2]   Department of Production Management, Bialystok University of Technology, Wiejska Street 45A,
      15-351 Bialystok, Poland; a.wasiak@pb.edu.pl
[3]   Faculty of Mechanical Engineering, Department of Production Computerization and Robotization,
      Lublin University of Technology, Nadbystrzycka 38 D, 20-618 Lublin, Poland; a.swic@pollub.pl
*   Correspondence: karol_tucki@sggw.pl (K.T.); o.orynycz@pb.edu.pl (O.O.); Tel.: +48-593-45-78 (K.T.);
      +48-746-98-40 (O.O.)

Received: 27 November 2019; Accepted: 21 January 2020; Published: 23 January 2020

Abstract: Managing of wastes rich in lignocellulose creates the opportunity to produce biofuels that are in full compliance with the principles of sustainable development. Biomass, as a suitable base for the production of biofuels, does not have to be standardized, and its only important feature is the appropriate content of lignocellulose, which assures great freedom in the selection of input. Biobutanol, obtained from this type of biomass, can be used as fuel for internal combustion engines, including diesel engines. In the era of strict environmental protection regulations, especially concerning atmospheric air, any new fuel, apart from good energetic properties, should also show beneficial ecological effects. This study investigates the carbon dioxide emissions from biobutanol powered diesel engine by means of use of the simulation model. The parameters of a real passenger car powered by a diesel engine were used for simulation carried out accordingly to the WLTP (Worldwide Harmonized Light Vehicle Test Procedure) approval procedure as the current test for newly manufactured cars. The results obtained for biobutanol were compared with simulated exhaust emissions obtained for conventional diesel and with FAME (fatty acid methyl esters)—the most popular biofuel. Biobutanol, in spite of its higher consumption, showed lower direct carbon dioxide emissions than both: the conventional diesel and FAME. In addition, a LCA (life cycle assessment) was carried out for the fuels and vehicles in question using the SimaPro package. Therefore, the implementation of butyl alcohol as a fuel provides favorable environmental effects. This result gives arguments towards biofuel production management indicating that implementation of biobutanol production technology mitigates carbon dioxide emission, as well as promotes lignocellulosic resources rather than edible parts of the plants.

Keywords: biobutanol; clean combustion; Scilab simulations; SimaPro; CO_2 emission; fuel production management; environmental impact; non-edible resources for biofuel production

---

1. Introduction

Due to the vast possibilities of obtaining useful products, utilization of lignocellulose rich waste is being analyzed in the area of natural and technical sciences with growing frequency [1,2]. In the period of deteriorating environmental condition, it is essential that the principles of sustainable development are followed in virtually all sectors of the economy in order to maintain balance between economic growth and concern about the nature around us [3,4]. Progressing climate changes, strictly connected with emission of anthropogenic origin greenhouse gases, constitute an important factor fostering search for state-of-the-art technologies, in particulars in the energy and transport industries [5–7]. Research into technologies to increase the energy efficiency of sustainable energy technologies for transport is gradually accelerating. It is important to create greater synergies and consistency between policies, as well as to develop a favorable regulatory, financial, and social environment. In addition, it should be based on global standards, processes and tools to manage safety, environmental protection, and cooperation with local communities.

As part of integrated efforts aimed at climate protection, in the global agreement—the Paris Agreement—the European Union undertook to maintain average global temperature growth on the level of 1.5 °C as compared with the pre-industrial period by limiting carbon dioxide emissions from its area by 40% until 2030 (as compared with 1990) [8,9]. Road transport which, thanks to numerous advantages, is developing dynamically and thus increasing its share in burdening the environment is becoming an increasingly important source of negative emissions [10–13]. To reduce the consumption of conventional sources as energy media for vehicles, biofuels, and bio-additives are applied, among others in order to replace exhaustible fuels emitting significant quantities of carbon dioxide [14,15]. Pursuant to Directive (EU) 2018/2001 of the European Parliament and of the Council of 11 December 2018 on the promotion of the use of energy from renewable sources, pure vegetal oils, methyl and ethyl esters of fatty or animal acids, mixtures of esters with diesel oil, and co-processed oils may be defined as biofuels suitable for diesel engines [16]. The fuels identified in the Directive, which may be used in the case of diesel engines also include biomass derived alcohols, such as: methanol, ethanol, propanol, and butanol [17–19]. Ethanol and methanol are most commonly used as alternative diesel engine fuel, but their different physical and chemical properties, such as: cetane number, flash point, low lubricity, low calorific value, poor miscibility with other substances and high volatility, materially hinder their suitability as diesel engine fuels [20,21]. Butanol seems an interesting alternative, due to its lower flash point, lower volatility, higher calorific value and cetane number, as well as better lubricity than methanol and ethanol [22,23]. What is more, butanol is characterized with better miscibility with conventional diesel fuel, vegetal oils, fatty acid methyl esters (FAME), and lower corrosivity [24–26].

Butanol, or butyl alcohol, is a colorless flammable substance produced in the process of anaerobic fermentation of sugar rich matter with the use of Clostridium bacteria, or from solid fuels [27–29]. It is a good solvent of petroleum derivative substances, while demonstrating low water solubility [30,31]. It appears as four isomers (iso-butanol, tert-butanol, sec-butanol, n-butanol), differing in terms of hydrocarbon chain structure (branched vs. straight) and hydroxyl group location [32,33]. Butanol is commonly used as paint and lacquer solvent, and as an ingredient of hydraulic oils and brake fluids [34,35]. It finds numerous applications in the textile and cosmetic industries. n-butanol is most frequently used in research on engines, as an additive or independent diesel fuel, because of its favorable physical and chemical properties (enumerated in the preceding paragraph) [36,37]. Butyl alcohol obtained from biomass is referred to as biobutanol [38,39]. Biobutanol may be produced from such plants as sugarcane and sugar beet, corn, grains, as well as derivative organic products obtained from agriculture and forestry, including straw, plant stalks, or wood waste [40–43].

The comprehensive research results presented in the literature aimed to compare the properties of butanol with those of conventional gasoline and diesel fuel as well as widely used biofuels—i.e., methanol, ethanol, and biodiesel—indicate that butanol has the potential to overcome many aspects of the disadvantages of low-carbon alcohols [44–46]. The main advantages of butanol include lower volatility, lower ignition problems, good mutual solubility with diesel without any solvents, more suitable viscosity as a substitute for diesel, and higher heating value [47].

The calorific value of butanol is higher (33.1 MJ/kg) as compared to ethanol (24.8 MJ/kg) and methanol (22.7 MJ/kg). This parameter, in combination with a higher stoichiometric air–fuel ratio, allows the use of higher levels of its share in motor gasoline without changing the engine control systems and distribution network. Oxygen content can improve the combustion process, resulting in less carbon monoxide (CO) and hydrocarbons (HC) emissions. In addition, butanol has a lower heat of vaporization than ethanol, which can reduce problems with fuel atomization and combustion in cold engine start conditions compared to typical ethanol fuels [48,49].

Compared with biodiesel, butanol contains more oxygen, which can reduce soot emissions, and nitric oxide (NO_x) emissions can also be reduced due to higher heat of evaporation, resulting in a lower combustion temperature.

Among the major disadvantages of n-butanol should be noted the increased fuel flow due to the lower calorific value compared to gasoline (44.5 MJ/kg) or diesel (44 MJ/kg). In addition, a lower octane number than in the case of low-carbon alcohols inhibits the use of a higher compression ratio in higher efficiency spark ignition engines—and also, as a gasoline substitute—may create a potential problem due to higher viscosity [50].

With respect to the impact of butanol on CO, THC (total hydrocarbon), and NO_x emissions, it should be emphasized that they can be reduced or increased depending on the specific engine (i.e., with partial or direct injection), operating conditions (i.e., with or without control of the air-fuel ratio, type of the timing gear), and on the mixture ratio [51].

The results of laboratory investigations on the impact of mixtures of biobutanol with diesel fuel on the combustion and emission characteristics of a four-cylinder diesel engine are presented in [52]. The tested fuels were a mixture of 10% biobutanol and 90% conventional diesel, 20% biobutanol and 80% diesel, and 100% diesel based on weight. The measurements were made at an engine speed of 1500 rpm and 30 Nm and 60 Nm engine load. NO_x, CO, and soot emissions were lower than those from diesel under all test conditions, while HC emissions were higher than from diesel.

In addition, as the content of butanol in the mixed fuel increased, the experimental results showed that the ignition delay was longer than the ignition delay for diesel fuel for all studied injection times. The indicated unit fuel consumption of mixed fuels was higher than diesel fuel consumption. However, the exhaust gas temperature was lower than those from diesel fuel at all injection times.

Additionally, the results of studies on the CO, CO_2, THC, NO_x emissions for various types of cars as a function of fuel composition (e.g., butanol share) are presented in [53,54].

The CO_2 emission values resulting from the use of n-butanol mixtures in diesel [55] and gasoline [56] engines in the context of the NEDC (New European Drive Cycle) driving test decrease, depending on the share of butanol in the mixture, by about 1–6% in comparison to pure gasoline or diesel (e.g., The mixing of 20% n-butanol with gasoline reduced CO_2 emissions by 5.7%).

Energy recovery of lignocellulosic waste material in the form of liquid fractions can yield alcohol-based fuels such as bioethanol or biobutanol [57].

Physicochemical properties of fuels and their mixtures influencing chemical reactions in the combustion process and, consequently, gas emissions are discussed in detail in [58,59]. When analyzing biofuels, particular attention is paid to autoignition reactions and the rate of heat release [60,61].

Although butanol properties (boiling point, viscosity, octane number) predetermine it for the use in spark ignition engines as a partial substitute for conventional gasoline, a number of studies were carried out using butanol/diesel fuel mixtures in compression ignition engines.

The advantage of butanol is its ability to reduce the viscosity of composite fuels, especially when mixed with FAME or crude vegetable oil [62–65].

The results of laboratory tests aimed at determining the impact of a mixture of butanol derived from lignocellulosic material and FAME based on animal fat on specific fuel consumption and CO_2, CO, NO, HC, and PM emissions of a diesel engine are available in the literature [62,66].

In [62], biobutanol derived from lignocellulose material was tested, which was then used as an additive for diesel engines. Biobutanol was used in fuel mixtures with FAME in the amount of 10%, 30%, and 50% butanol. 100% diesel and 100% FAME were used as reference fuel.

The laboratory tests carried out showed that the use of biobutanol in fuel reduced the production of carbon dioxide (by 15%), nitrogen oxides (by 35%), and PM (by 90%). Moreover, the use of biobutanol as an additive in FAME, especially in oils, significantly reduced the viscosity and density of the fuel.

Also in [67,68], conventional diesel, 30% biodiesel (FAME) and biodiesel with 25% n-butanol in a turbocharged diesel engine were compared. In all cases, the positive effect of butanol in diesel fuel on particulate, NOx and carbon dioxide emissions was found. In addition, a positive effect on smoke emissions has been noted for the n-butanol mixture.

The main disadvantages of FAME include poorer storage and oxidative stability. The high cost of the raw material is also important, especially when using vegetable oil as a raw material [21].

For environmental, logistic, and economic reasons, lignocellulosic biomass is a particularly attractive raw material for biofuel production. When selecting appropriate conversion methods, it is possible to obtain, among others, cellulosic ethanol, synthetic gas (bio-SG), or increasingly appreciated furan fuels (Furanics fuel) [69–71]. Furan fuels, i.e., compounds derived from furan, have been identified in the "Roadmap for Biofuels in Transport" prepared by the International Energy Agency (IEA) as future biofuels for which intensive development of production technology is expected by 2050. Currently, work on technologies for obtaining them is only at the stage of research and development.

Lignocellulose rich wood waste is very attractive thanks to a non-food character of the substrate and possibility to obtain second generation biofuel, i.e., fuel not competitive towards food. Poland has significant potential for production of waste wood and wood residues. Towards the end of 2017, the area of Polish forests amounted to 9242 thousand ha, which corresponds with the forest coverage ratio of 29.6% [72,73]. To compare, at the end of 2010 the area of forests was 9121.3 ha, which was equivalent with the forest coverage ratio of 29.2%. Thus, an increase in the country's forest coverage ratio by 0.4% was recorded [74,75]. In 2017, 42,699 thousand m^3 net of thick wood was produced in Poland, including 8607 thousand m^3, i.e., 21.2% of all thick wood volume obtained in connection with forest clearing, acquisition of deadwood, wind broken trees, and trees damaged as a result of various weather occurrences and natural processes [76]. Thick wood production of 31,822 m^3 was recorded in 2010, including 5686 m^3 (17.8% of all thick wood obtained) of wood from forest clearing and ordering processes [77]. The figures do not include information on clearing of greens located along roads, acquisition of wood stock from parks and city green areas, residue from sawmills and wood processing companies; therefore, the quantity of lignocellulose rich matter which may be used for biofuel production is in fact much higher [78]. The choices of directions of technology development, as well as the choices of biofuel production technology are the matter of economic, social, political, and environmental issues. The carbon dioxide emission is one of the factors determining usefulness of particular biofuel production technology. Moreover, the use of edible or nonedible resources is another factor strongly affecting the eventual choice of technology. Consequently, the knowledge and understanding of phenomena occurring during burning of various fuels (including biofuels) in automotive, as well as other types of internal combustion engines, assures possibility of decisions concerning choices of types of biofuels and technologies of their production. All these factors, when established, provide tools for technology management in the area of biofuel production and distribution. This technological knowledge is also needed for undertaking legal decisions concerning allowable content of fuels available on the market.

The life cycle assessment methodology (LCA) [79,80] is increasingly used to maintain the environmental sustainability of biofuels [81,82].

The LCA investigations should be performed at the stage of technological process designing, what could result in more effective controls of environmental issues [83,84].

There are several recognized methods for assessing the life cycle impact, e.g., EPS 2000 (priority strategies for environmental protection), CML (Centrum voor Milieuwetenschappen Leiden), ecological indicator 99, IMPACT 2002+, ReCiPe, MIPS (material intensity per service) [85–90]. They are implemented in computer programs used in the LCA investigations, eg SimaPro, GaBi, Umberto [91–93]. Based on the results of computer calculations, under specific assumptions, it is possible to estimate the environmental impact of selected products or production processes. Comparing the results obtained, it is possible to indicate a product or production process that will have a minimal impact on the environment.

2. Materials and Methods

Taking into account the potential of obtaining lignocellulosic biomass which may be used for production of biobutanol, properties similar to conventional fuel and mitigation of negative environmental impact, this paper focuses on analyzing the impact of butyl alcohol on diesel engine carbon dioxide emissions.

The test used a simulation model of a contemporary diesel engine, reflecting the predefined work cycle. This allows elimination of hazard to the actual engine which, at the development stage, did not consider being powered with fuel different than conventional [94–96], while enabling achievement of complete results in an accessible form and within a short time.

2.1. Materials

The simulation used actual parameters of the Fiat Panda passenger car with a modern diesel 1.3 MultiJet II drive unit, compliant with the Euro 6 norm for exhaust gases [97–99]. The vehicle is equipped with the "start–stop" system, aimed at reducing the quantity of consumed fuel and the quantity of emitted exhaust gases [100,101]. The system allows switching the engine off if its operation is not needed at the given moment. In urban traffic, this takes place when the vehicle is not moving, which happens frequently due to the infrastructure and traffic management system. Such solutions enable reduction of exhaust gases and noise which, as fuel is not combusted and the drive unit is not in operation, are not emitted. Development of the engine stopping system while the vehicle is not in motion is caused by increasingly stringent limits governing harmful substance emissions and sound levels. Continuously decreasing acceptable values force engineers to seek additional solutions next to mere engineering modifications in contemporary combustion engines.

Emission limits for newly manufactured vehicles have been in operation since 2009 (the first application of emission limits in 2015). The limits on average carbon dioxide emissions for passenger cars in the European Union will be regular reduced from 130 gCO$_2$/km in 2015 to 65 gCO$_2$/km in 2030 [102–104].

The analyzed vehicle is characterized with the maximum power of 75 hp at 4000 rev./min and the maximum torque of 190 Nm at 1500 rev./min. Pursuant to the manufacturer's data, fuel consumption in the urban cycle is 4.7 l/100 km, in the mixed cycle 3.9 l/100 km, with 3.5 l/100 km outside the city [105].

Table 1 presents basic technical parameters of the engine used in the simulation.

Table 1. Basic technical data of the engine 1.3 MultiJet II used in the simulation.

Parameter	Unit	MultiJet II
Cylinder layout	-	in-line
Number of cylinders, c	-	4
Type of injection	-	direct, multistage
Compression ratio, e	-	16.8:1
Diameter of the cylinder, D	mm	69.6
Piston stroke, S	mm	82
Engine displacement, Vss	cm^3	1251
Maximum engine power, Ne	kW	55
Engine rotational speed for its maximum power, nN	rpm	4000
Maximum engine torque, Me	Nm	190
Engine rotational speed for its maximum torque, nM	rpm	1500
Rotational speed of idle gear, nbj	rpm	850 ± 20

Table 1 gives the factory characteristics of the engine (selected parameters were used in the developed computer simulation) assuming that it is fully technically sound. Based on the characteristics given in the table, the full course of CO$_2$ emissions cannot be determined for changes in load and engine speed. However, additional charts presented in the part of the article "The simulation model" show the course of CO$_2$ changes used for the simulation as a function of rotational speed and engine load for the assumed fuels. These characteristics take into account the chemical composition of the fuel and calorific value, which determine the amount of fuel consumed and, consequently, the amount of CO_2 emitted.

The simulation model used in the experiment enables utilization of fuels characterized with different properties. In order to analyze the emission profile, parameters of the following fuels were implemented: conventional diesel oil ON (as reference fuel), fatty acid methyl esters (FAME) (as the most common diesel fuel bio-additive), and butanol (main subject of the analysis). These fuels indisputably differ in terms of elementary composition and properties necessary from the point of view of combustion of the respective fuel. Table 2 compares properties of the fuels used.

Table 2. Selected properties of the fuels applied. FAME—fatty acid methyl esters.

Parameter	Diesel	FAME	Butanol
Carbon content (%)	86.5	78.0	64.8
Hydrogen content (%)	13.4	12.0	13.5
Oxygen content (%)	0.0	10.0	21.6
Calorific value (kJ/g)	44.0	37.0	33.1
Air demand (g_{air}/g_{fuel})	14.5	12.5	11.2

The data in Table 2 demonstrate differences among the properties of diesel and alternative fuels. The greatest difference is noticeable in terms of oxygen content, which is absent in classic fuel, but present in biofuels, as well as in terms of carbon content where diesel shows the highest level among the analyzed fuels. Moreover, conventional fuel is characterized with the highest calorific value [106–110].

From the point of view of physicochemical properties and elementary composition, the use of 100% biofuel is not possible; however, highly desirable due to the ecological effects. The vehicle under analysis is certainly not 100% suitable for alternative fuels. Therefore, the authors used computer simulations in their research. The simulation considered both pure fuels and mixtures with diesel.

Table 3 presents particular proportions of fuels used in the model.

Table 3. Proportions of fuels used in the model.

Diesel/FAME	Diesel/Butanol
100%/0%	100%/0%
90%/10%	90%/10%
80%/20%	80%/20%
70%/30%	70%/30%
60%/40%	60%/40%
50%/50%	50%/50%
40%/60%	40%/60%
30%/70%	30%/70%
20%/80%	20%/80%
10%/90%	10%/90%
0%/100%	0%/100%

Moreover, the simulation takes into consideration operation of the "start–stop" system installed in the vehicle and, therefore, it reflects actual movement of the car in accordance with the predefined test.

2.2. Methods

2.2.1. Simulation Model

The simulation model used in the analysis was developed in the Scilab environment, i.e., free of charge scientific software enabling execution of advanced mathematical calculations and algorithms [111]. It allows designing, performance of simulations, combining and recording of projects. Thanks to the possibility to resolve differential equations, linear and nonlinear systems, application of fast Fourier transform, development and optimization of algorithms, is an extremely useful tool in the case of more complex systems [112,113]. The Xcos package was used in connection with this analysis to prepare block diagrams reflecting actual dependencies in the analyzed engine. Graphical presentation of the modeled system, which is simple to use and which minimizes the risk of calculation error, is an indisputable advantage.

To be able to relate to reference emission levels applicable to the analyzed vehicle, the simulation used the WLTP procedure, being the latest homologation test which has covered all newly manufactured vehicles applying for traffic approval in the European Union since 1 September 2018 [114,115]. The previous test procedure was developed in the 1980s and it based on a theoretical driving profile. At the beginning, could be considered reliable, but dynamic technological development may provide additional variables which were not considered therein. A generalized approach to the model driving cycle and identical treatment of all analyzed vehicles rendered numerous irregularities and discrepancies revealed during more detailed tests of specific cars. It therefore became necessary to amend the vehicle testing procedure for cars to be launched on the market [116]. By assumption, the WLTP test reflects the actual vehicle operating conditions, considers the equipment installed in the vehicle, engine versions, as well as gear settings. The new WLTP test cycle takes 10 minutes longer than the previous procedure and the vehicle covers a distance by 12.25 km longer than during the NEDC test. Moreover, the new procedure features as many as four dynamic work phases, with approximately 52% in the urban cycle and the remaining 48% reflecting driving outside the city. The temperature in which testing is performed is also important. In the case of WLTP, it is the range of 14–23 °C, whereas NEDC was carried out in the range of 20–30 °C, which was remote from actual European conditions [106,117,118].

The diagram of the simulation model applied in the test is presented in Figure 1.

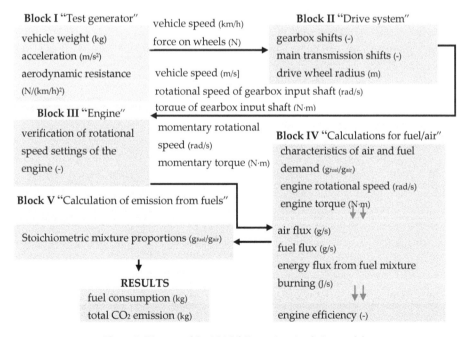

Figure 1. Diagram of the 1.3 MultiJet engine simulation model.

The simulation was divided into five blocks, with each one performing a separate function and providing vehicle operation and fuel combustion parameters necessary to carry out the WLTP test procedure and obtain results related to carbon dioxide emission for the analyzed fuels. Presented below.

Block I "Test generator"

The block is responsible for furnishing correct parameters, characteristic for the WLTP driving test, the distance traveled by the vehicle l, including vehicle acceleration a(t), vehicle speed and force generated on the wheels P_1 [N] (calculation weight of 1020 kg). In order to determine these values, the module takes advantage of the vehicle's technical data, such as: vehicle weight, rolling resistance, and aerodynamic resistance.

$$1 = \int_0^t v(t)dt[m] \tag{1}$$

$$a(t) = \frac{dv(t)}{dt}\left[\frac{m}{s^2}\right] \tag{2}$$

where:

v(t)—the momentary speed of the vehicle in the test (m/s)
t—the end time of simulation (s).

$$P1 = F0 + F1 \cdot v(t) + F2 \cdot v(t)^2 \ [N] \tag{3}$$

where:

$v(t)$—the momentary speed of the vehicle in the test (km/h)
$F0$—rolling resistance coefficient (N)
$F1$—linear resistance coefficient (km/h)
$F2$—aerodynamic resistance coefficient (N/(km/h)^2).

The vehicle loads resulting from the accelerations acting on it are then calculated using equation:

$$P2 = M\frac{dv(t)}{dt} \ [N] \tag{4}$$

where:

$v(t)$—the momentary speed of the vehicle in the test (m/s)
M—vehicle calculation weight (kg).

Table 4 presents values adopted for the analyzed vehicle.

Table 4. Vehicle parameters implemented in the "test generator" block.

Parameter	Unit	Fiat Panda
Vehicle weight	kg	1020
Rolling resistance	N	6.15
Aerodynamic resistance	N/(km/h)^2	0.0412

Figure 2 below presents results obtained from that simulation block. The graphs illustrate such values as: v—speed of the analyzed vehicle; d—road in kilometers, covered during the test; F—force acting on the vehicle's wheels; and p—current gear in which the vehicle is driving.

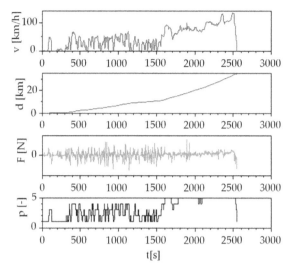

Figure 2. Parameters obtained from the block responsible for generating the WLTP test signal.

Block II "Drive system"

This block involves calculations related to such signals as the vehicle driving speed, gearbox input shaft torque, its rotational speed for the gearbox. In order to determine these values for the model correctly, data regarding the vehicle wheel radius, drive system ratios for the rotational speed, and drive system ratio for torque were implemented.

$$Mun = (P1 + P2) \cdot R \cdot R1 \ [N \cdot m] \tag{5}$$

where:

Mun—torque acting on the gear shaft (N·m)
$P1$—momentary force on wheels from resistance to motion (N)
$P2$—momentary force on wheels from inertia (N)
R—wheel radius (m)
$R1$—drive system shifts for the torque (-).

$$wun = \frac{v(t)}{R} \cdot R2 \left[\frac{rad}{s} \right] \tag{6}$$

where:

wun—rotational speed of the gear shaft (rads/s)
$v(t)$—the momentary speed of the vehicle in the test (m/s)
R—wheel radius (m)
$R2$—drive system shifts for rotational speed (-).

Table 5 identifies the values of those parameters used in the developed model.

Table 5. Parameters for the 'drive system' module based on the manufacturer's data.

Parameter	Unit	Fiat Panda
Radius of the vehicle wheels	m	0.298
Drive system ratio for torque	(-)	0; 1/13.46; 1/7.05; 1/4.55; 1/3.24; 1/2.42
Drive system ratio for rotational speed	(-)	0; 13.46; 7.05; 4.55; 3,24; 2.42

Results obtained from this block are presented in Figure 3. The graphs correspond, respectively, for: ωk—vehicle wheel angular velocity; ωp—engine angular velocity; Mk—torque on vehicle wheels; and Mp—torque on engine.

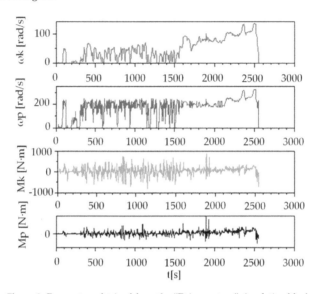

Figure 3. Parameters obtained from the "Drive system" simulation block.

Block III "Engine"

The "Engine" module plays a verification role in the developed model. It is responsible for verification of values obtained from the preceding blocks in terms of presence thereof in the admissible range of engine rotational speed variability. This allows eliminating deviations and distortions in the model, which could lead to incorrect results. If previously obtained rotational speed momentary values (rad/s) and torque values (N·m) are correct, they are passed on to further simulation blocks.

$$Msi = \begin{cases} Mun; Mun > 0 \, \text{N} \cdot \text{m} \\ 0 \, \text{N} \cdot m; \, Mun \leq 0 \, \text{N} \cdot \text{m} \end{cases} \, [\text{N} \cdot \text{m}] \tag{7}$$

$$wsi = \begin{cases} wun; wun > 83.7 \, \text{rad}/\text{s} \\ 83.7 \frac{rad}{s}; \, wun \leq 83.7 \, \text{rad}/\text{s} \end{cases} \left[\frac{rad}{s} \right] \tag{8}$$

where:

Msi—momentary torque on the vehicle's gear (N·m)
ωsi—momentary rotational speed on the vehicle's gear (rad/s).

Block IV "Calculations for fuel/air"

"Calculations for fuel/air" is the most structurally developed block, which allows determination of fuel (Figure 4) and air consumption values (Figure 5), necessary for conducting correct fuel combustion in the test and determination of aggregated values of those elements. The module is based on universal fuel and air mixture demand characteristics.

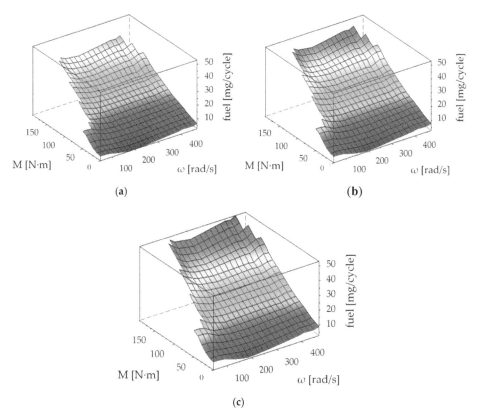

Figure 4. Fuel flow per injection cycle: (**a**)—Diesel; (**b**)—FAME; (**c**)—Butanol.

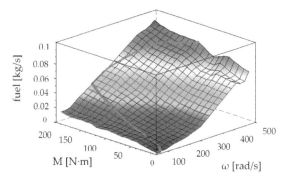

Figure 5. Characteristics of the hourly fuel consumption as a function of the rotational speed and torque of the selected internal combustion engine adopted in the simulation.

The characteristics used in the simulation were developed on the basis of published experimental studies conducted on the engine test bench. Detailed numerical values from which the spatial distributions were based (Figures 4 and 5) present literature items [98,99].

Figure 4 shows the characteristics of the output of the analyzed fuels per one diesel engine injection as a function of changes in speed and changes in torque produced by the engine.

Diesel fuel is characterized by the lowest values of fuel expenditure per injection as a function of engine speed and torque generated by the engine due to the highest calorific value among the analyzed objects (44 MJ/kg). For the fuel with the lowest calorific value of butanol (33.1 MJ/kg), the characteristic curve of output per fuel cycle for changes of rotational speed and torque takes the highest values. In the case of FAME fuel with calorific value (37 MJ/kg), the fuel flow rate per 1 injection is obtained between fuels with extreme calorific values.

It is seen on Figure 4 that plots shown on the figures are very similar with respect to their shape, only the values of fuel flow at corresponding points are slightly different. The whole plot for butanol is located above the FAME, and diesel fuel occupies the lowest position.

Determination of values related to the fuel flux and required quantity of air, taking into account the fuel calorific value, produces the value of energy flux from burned fuel. That is followed by determination of temporary engine efficiency. With all of the above parameters, based on integrating modules, one can determine values for the fuel and air mixture consumed in the test.

$$\dot{fuel} = ffuel(\omega si, \ Msi) \left[\frac{g}{s} \right] \tag{9}$$

$$\dot{air} = fair(\omega si, \ Msi) \left[\frac{g}{s} \right] \tag{10}$$

where:

$\dot{fuel}$—fuel flux (g/s)

ffuel—function of hourly fuel consumption depending on rotational speed and torque (g/s)

$\dot{air}$—air flux (g/s)

fair—function of hourly air consumption depending on rotational speed and torque (g/s).

The simulation developed provides for the possibility to include or exclude simulation elements accounting for operation of the start–stop system. If the vehicle stops during the driving test simulation performed and the said system is switched on, calculations are performed in accordance with the following dependencies:

$$\dot{ons} = \begin{cases} \dot{on}; v(t) \ > 0 \ \text{m/s} \\ 0; \ v(t) = 0 \ \text{m/s} \end{cases} \left[\frac{g}{s} \right] \tag{11}$$

$$\dot{airs} = \begin{cases} \dot{air}; \ v(t) \ > 0 \ \text{m/s} \\ 0; \ v(t) = 0 \ \text{m/s} \end{cases} \left[\frac{g}{s} \right] \tag{12}$$

Figure 6 presents graphs of such results as: fuel—fuel consumption; air—air consumption; Fuel—aggregated value of fuel consumed in the test; Air—aggregated value of air consumed in the test.

Block V "Calculation emission from fuels"

The block is responsible for determination of carbon dioxide emission value for the respective fuel. Carbon dioxide emissions per diesel, FAME, and biobutanol injection cycle are shown in Figure 7.

For carbon dioxide emissions per diesel injection cycle, as shown in the diagrams below, there are clearly smaller differences in the characteristics of the speed and torque variations than for the fuel expenditure characteristics shown above. This may be due to the high oxygen content of biofuels that have already been chemically bonded to carbon atoms, which results in a slight increase in carbon dioxide emissions.

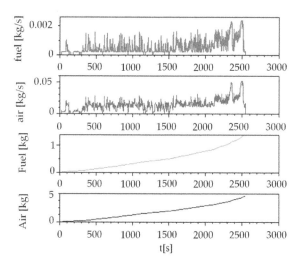

Figure 6. Parameters obtained from the "Calculations for fuel/air" simulation block.

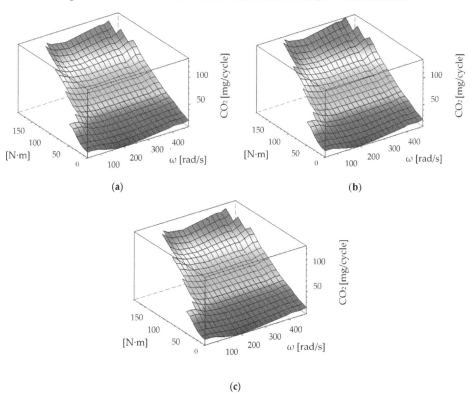

Figure 7. Carbon dioxide emissions per diesel, FAME, and biobutanol for one injection cycle: (**a**)—Diesel; (**b**)—FAME; (**c**)—Butanol.

Using the engine emission profile, it defines momentary CO_2 mass values and aggregated values throughout the test.

$$\dot{CO_2} = fCO_2(ws, \ Ms) \ gCO_2 \ \dot{spal} / gspal \left[\frac{g}{s} \right] \tag{13}$$

$$CO_2(t) = \int_0^t \dot{CO_2}(t) \ dt \ [\text{g}] \tag{14}$$

2.2.2. LCA Method

The LCA methodology is exists in several variants and interpretations of key concepts. Thereare two alternative approaches used in LCA—the attributional model and the consequential model. Attributional life cycle assessment (ALCA) estimates the share of global environmental loads that a product belongs to. The consequential LCA (CLCA) gives an estimate of how global environmental loads affect the product production and use. The distinction was created to resolve debates on what input should be used in the LCA and how to deal with allocation problems. ALCA is based on average data, and the allocation is done by sharing the environmental burden of the process between the life cycles supported by this process. CLCA in principle uses marginal data in many parts of the life cycle and avoids allocation through system expansion.

Each of the models is associated with significant parametric and model uncertainty, and estimating the impact of biofuels on the climate requires many subjective choices [119–121].

A review of the literature for biofuel modeling shows that the authors use both CLCA [122–125] and ALCA method [126,127].

It should be emphasized that the intention of the authors of the present paper is not to attempt to disqualify the specific capabilities of any of the LCA analysis models, but to attempt to analyze the behavior of a non-steady state complex system.

The simulation model developed in connection with this analysis provides data on direct carbon dioxide emissions. Emissions originate from elementary composition of the fuels used and characteristics of the combustion process in the respective engine [128,129]. Results are obtained based on differential equations, characterizing the process of fuel burned by the vehicle and resulting quantities of exhaust gases emitted from the exhaust system. The method is in line with vehicle homologation tests; yet, in the context of environmental impact assessment for particular fuels, it may be insufficient. To supplement the analysis, it was extended to cover the LCA analysis, which is considered crucial for accomplishment of the sustainable development policy and a reliable tool for conducting environmental analyses in the context of specific products, including fuels. LCA tools are perceived as a foundation of state-of-the-art management of the environment related and decision processes which have real impact onto various areas of natural economy. Due to their scope and a broader perspective of the respective product, numerous state strategic documents and policies identify LCA analyses as mandatory. The product life cycle considers consumption of materials, energy, and resources, as well as the effects of processing thereof not only at the time of actual use of the product, but also at the stages of production and disposal. The analysis commences from mining of resources necessary for manufacturing the product and includes all energy and material expenditures connected therewith. Subsequently, it focuses on manufacturing of a specific product. Next, it considers the stages connected with using it, to finally take into account the process of disposal or decomposition. It can therefore be seen that it is much more detailed than typical analyses focusing solely on direct use of the item in question. LCA analyses prove extremely useful during decision making processes based on identification of processes or products which will be the least harmful to the environment throughout their life cycle. In combination with knowledge regarding costs, ease of use, and production technologies, one may identify solutions constituting the least burden to the natural environment, which is strictly connected with management in accordance with the ideas of sustainable development and performed more and more often worldwide. Detailed techniques related to process and product life cycle assessment are defined in the ISO 14040 international standard [130]. The document defines the necessary documents required for proper execution of the analysis, including: inventory of the set of material data; environmental impact analysis of elements connected with the identified data; interpretation of results of performed analysis, as well as reference of the impact assessment to the research subjects analyzed.

Life cycle assessment ought to contain an identified objective and scope of analysis, identified data set, assessment of impact onto particular elements, and properly interpreted results.

The LCA analysis of the fuels used was performed in three stages. It was accomplished with the SimaPro ver. 9.0.0.48 software, designed for execution of professional environmental impact analyses both in business and scientific areas. The tool enables the analysis and monitoring of important information from the sustainable development perspective. With the use of that environment, one may perform modeling and analysis of even complex product and process life cycles as well as of their actual environmental impact on each of the stages. What is more, SimaPro is fully compliant with the guidelines identified in the ISO 14040 standard and, as such, it constitutes a source of reliable results which may be used in product related decision-making processes.

The first stage of the analysis involved determination of the impact from production of the fuels in question onto particular elements of the environment. It considered the following areas: "ecosystem quality", "climate change", "human health", and "resources" in the context of carbon dioxide emission during fuel unit production processes. This stage involved use of the IMPACT 2002+ method, which identifies impact of the analyzed product onto the environment and people. The method bases on modern exotoxicity comparative analysis for both environmental elements and those related to human health.

The next stage of analysis involved the Greenhouse Gas Protocol method, based on which information regarding carbon dioxide emission is obtained. The method is based on the greenhouse gas emission protocol and distinguishes four result segregation categories: "Fossil CO_2 eq"—carbon dioxide emission from fossil fuel conversion; "Biogenic CO_2 eq"—emission caused by plants and trees; "CO_2 eq from land transformation"—emission connected with transformation of land; and "CO_2 uptake"—that is carbon dioxide value captured during the given process. The analysis relates to the production stage of a specific product. Any information concerning the methods of formation, energy consumption during the processes, and their progress is contained in extensive libraries of the SimaPro software.

The final, third stage utilizes data obtained pursuant to the Greenhouse Gas Protocol method regarding carbon dioxide emission during production of a reference unit of the analyzed fuel. Emission values from the production stage are combined with data obtained from the simulation model, which allowed assessment of the emission rate from the fuels throughout their life cycle, from production to conversion into thermal energy. Table 6 below presents input parameters applied in the LCA analysis of the discussed fuels.

The service of life car was adopted according to the LCA analysis performed for its cars by a leading manufacturer of commercial vehicles [131,132]. In addition, a literature review confirms that the value of 150,000 km is in accordance with ISO 14044 [110,130].

A relatively frequent practice in scientific publications is use of reference emission values published by vehicle manufacturers, which results in generalization of analysis results. One needs to stress that such a value relates solely to powering the vehicle with conventional fuel and, as such, it cannot be taken into consideration in analyses related to alternative fuels. It was, therefore, reasonable to implement in this study direct carbon dioxide results obtained from the developed computer simulation.

Table 6. Parameters used in LCA analysis.

Fuel Consumption in the Cycle (l/100 km)		
Fiat Panda	Urban	Extra urban	Combined
MultiJet II	4.7	3.5	3.9

Estimated Fuel Consumption in Car Service Life	
Diesel	6750
FAME	8250
Butanol	8250

CO_2 Emission from Fuels in Production Process (kgCO_2/1 kg fuel)	
Diesel	0.341
Butanol	3.18
FAME	−7.01

CO_2 Emission from Fuels in Burn Process (Start–Stop System ON) (g/km)	
Diesel	161,59
Butanol	173,16
FAME	161,33

CO_2 Emission from Fuels in Burn Process (Start–Stop System OFF) (g/km)	
Diesel	163,65
Butanol	175,39
FAME	163,39

Service life of car (km)	
150,000	km

3. Results and Discussion

3.1. Result from Simulation

The developed simulation allowed execution of an experiment using different mixtures of fuels with conventional diesel. Figure 8 below presents fuel consumption (value of the end of fuel used during the entire WLTP test) in the function of respective additive content.

Graph A (Figure 8) presents fuel consumption during vehicle operation taking into account the start–stop system, while in graph B (Figure 8) the system is switched off. Stars mark butanol, while circles—FAME fuel. As can be seen, higher fuel consumption for both operation modes is shown by butanol. Moreover, one can generally notice higher consumption of the medium in graph b, which confirms correct operation of the model considering the "start–stop" subsystem.

Figure 9 presents carbon dioxide emission profile graphs in the function of additive content, taking into account operation of the "start–stop" system (final value of CO_2 emission during the entire WLTP test).

Pursuant to the above graph, higher carbon dioxide was demonstrated by the FAME fuel. Butanol, along with increasing share of it as an additive in the mixture, reduced carbon dioxide emission. Similarly to fuel consumption, the "start–stop" system (Figure 9) contributed to much lower CO_2 emission than in the case of engine operation without it.

Table 7 below presents emission results in the test. Emission corresponds with total weight of carbon dioxide emitted for particular fuels (100% fuel content) after execution of the test cycle. Moreover, the value was recalculated into a comparative unit and compared with the reference value defined in the regulation [70].

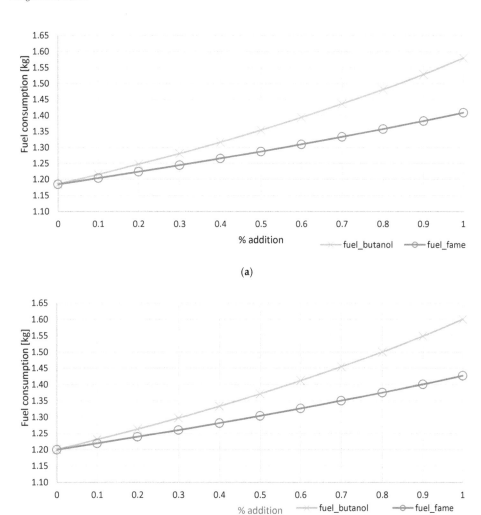

Figure 8. Fuel consumption as a function of the share of an additive ((**a**)—start–stop ON, (**b**)—start–stop OFF).

Table 7. List of CO_2 emission results for the individual fuels with values from regulations.

Fuel Used	CO_2 (kg) Emission	CO_2 (g/km) Emission	CO_2 (g/km) Value Required by the Regulation
		Start–Stop System ON	
FAME	4.026	175.48	130
Butanol	3.751	163.44	130
Diesel	3.757	163.87	130
		Start–Stop System OFF	
FAME	4.078	175.398	130
Butanol	3.799	163.398	130
Diesel	3.805	163.656	130

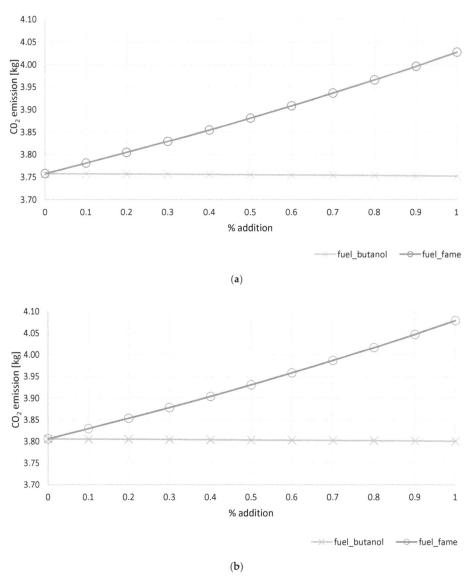

Figure 9. Carbon dioxide emission for fuel mixtures as a function of the share of an additive ((a)—start–stop ON, (b)—start–stop OFF).

The table above shows that analyzed biobutanol is characterized with the lowest total carbon dioxide emission. In terms of emission, conventional diesel fuel turned out second, while the most common biofuel (both as an additive and as fuel itself) was characterized with the highest emission level. This interesting result was verified on the basis of tests published in the literature, being carried out on an engine test bench. Studies confirm that carbon dioxide emissions at individual measuring points are highest when the engine is powered by methyl esters [133,134]. These results are due to the fact that rape oil fatty acid methyl esters (FAME) have a different elementary composition and different physicochemical properties influencing the course of processes occurring in the cylinder.

As seen in Table 8 the stoichiometric content of biobutanol and two examples of methyl esters of fatty acids the butanol has the smallest carbon content, and also the highest content of oxygen. These observations are in agreement with the emissions reported in Table 7.

Table 8. Stoichiometric content of biobutanol compared to two methyl esters of fatty acids.

Content %	C	H	O
Butanol	64.87	13.52	21.63
Methyl palmitate	75.56	12.6	11.86
Methyl stearate	76.52	12.76	10.74

The increase in carbon dioxide emission in the case of feeding the engine with plant oil esters is compensated for by the fact that, in this case, the carbon dioxide circulates in a closed circuit in the environment. Esters make a renewable fuel obtained from plants which, for production of organic matter in the photosynthesis process, use atmospheric carbon dioxide and release oxygen to the atmosphere.

The fuels in question exceed the admissible carbon dioxide emission level determined in the standard applicable to the specific vehicle. Still, one needs to emphasize that the analyzed vehicle was approved under the NEDC procedure which, as stated above, generalized the data related to movement of the vehicle in actual conditions.

3.2. Result from Simulation

Figure 10 presents analysis results for the analyzed fuels pursuant to the IMPACT 2002+ method. For the sake of graph legibility, presentation of results in the 'single score' mode, i.e., aggregated graph, was selected. In this form, results are presented in the mPt unit.

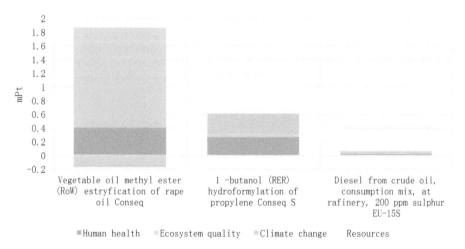

Figure 10. Impact of analyzed fuels onto selected environmental areas (mPt).

Based on Figure 10, it may be inferred that the fuel characterized with the greatest environmental impact at the production stage are fatty acid methyl esters (FAME). This solution demonstrates significant influence onto ecosystem quality, as the fuel is obtained from oil plants. Cultivation of such plants requires transformation of land in production related purposes, which results in a high value of the coefficient. However, this fuel has beneficial impact onto climate changes, similarly to biobutanol, because—as fuels obtained from organic matter—they have negative environmental impact thanks to natural carbon dioxide absorption. The above results from the life cycle of plants and trees they are obtained from, as an assumption is made regarding balancing of CO_2 absorbed during the photosynthesis cycle with carbon dioxide produced at further stages of the fuel's life. Conventional diesel fuel showed the greatest impact onto natural resources, which is caused by production of the fuel from crude oil.

The next step involved analysis related to production of selected fuels exclusively in terms of carbon dioxide emission. Figure 11 below presents results of the Greenhouse Gas Protocol method.

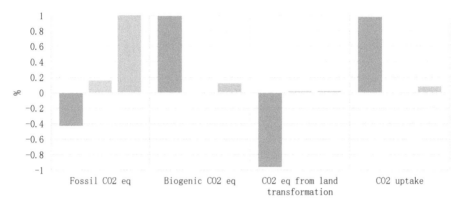

■ Vegetable oil methyl ester (RoW) estryfication of rape oil Conseq

▨ Diesel from crude oil, consumption mix, at rafinery, 200 ppm sulphur EU-15S

▨ 1 -butanol (RER) hydroformylation of propylene Conseq S

Figure 11. Particular carbon dioxide emissions occurring in production processes of selected fuels (%).

Based on the results obtained under the GGP method, it may be concluded that FAME is the fuel characterized with the lowest carbon dioxide emission in production processes. Butanol came second, while the highest level of emission was connected with production of diesel fuel. In the case of biofuels, negative emissions result from carbon dioxide absorption by plants used for production of the fuel. This process may also be included in production of butanol from waste biomass. Consequently, considering the whole product life cycle, it may be concluded that industrial fuel production is characterized by greater emissions than acquisition of energy media from organic matter.

The last stage of LCA analyses involved comparison of direct emissions obtained in the simulation with those obtained pursuant to the GGP method. In order to evaluate the whole life cycle of the fuels and their total emissions, the data were aggregated taking into account the assumptions stated in the Methods section. Results from combination of production emissions with those from fuel burning are presented in Table 9.

Table 9. Total carbon dioxide emission from analyzed fuels during life cycle.

	Diesel	FAME	Butanol
CO_2 total emission from fuels (start–stop system ON) (t)	26.98	−32.85	12.43
CO_2 total emission from fuels (start–stop system OFF) (t)	27.29	−32.51	12.74

Based on the above results, it may be concluded taking into account the whole life cycle of the fuel that the medium characterized with the lowest carbon dioxide emission turned out to be FAME. Analyzed biobutanol demonstrated an emission of 12.43 tonnes CO_2, which is still half of the emission from conventional diesel fuel characterized with the highest carbon dioxide emission level. In the comparison, influence of the start–stop system installed in the vehicle is also noticeable which, if used, resulted in the exhaust gas emission level lower by—on average—app. 1.66% than if the system was not installed in the vehicle. The greatest influence of the start–stop system was observable for biobutanol. The higher is hydrogen content in the fuel as compared to the carbon content, the lower is the CO_2 emissivity 13.4/86.5 (0.1549) ON, 12.0/78.0 (0.1538) FAME, 13.5/64.8 (20.83) biobutanol. It follows that diesel has the highest CO_2 emissions, followed by FAME, and biobutanol the lowest.

The results of the analysis of carbon dioxide emissions in the context of the fuel life cycle are shown in Figure 12.

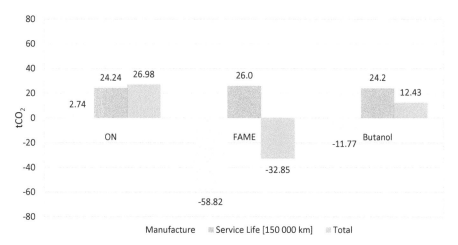

Figure 12. Comparison of carbon dioxide emissions from fuels in the context of the fuel life cycle.

Figure 12 shows that each fuel has different emissions depending on its life cycle. After taking into account the emission data resulting from the fuel life cycle, the final results are summarized in Table 10.

Table 10. Carbon dioxide emissions for a given fuel over the vehicle's life cycle.

Fuel	CO_2 Total (t)
Diesel	26.98
FAME	−32.85
Butanol	12.43

In the case of butanol, total emission is half of that recorded for conventional fuel.

The FAME fuel showed the lowest total emission, reaching more favorable values than both conventional fuel and butanol.

4. Conclusions

Analyses performed based on computer simulation allowed determination of the carbon dioxide emission profile for selected fuels, in accordance with the predefined test procedure and taking into account operation of additional systems in the vehicle ("start–stop" system).

- Biobutanol turned out to be characterized by the lowest emission levels. Interestingly enough, the trend maintained despite increased consumption of this fuel as compared with FAME. Fatty acid methyl esters demonstrated higher CO_2 emission than alcohol-based fuel, despite lower requirements for the medium.
- Simulation studies have confirmed that using the start–stop system, decreases CO_2 emissions and the consumption of the fuel mixtures investigated.

- Total carbon dioxide emission was lowest in the case of biobutanol (3.80 kg in the test), followed by diesel fuel (3.81 kg in the test), and highest for FAME (4.08 kg in the test). The CO_2 emission reduction obtained from the simulations is consistent with the results of the experimental studies referred to in the references review.
- As compared with the exhaust gas emission standard applicable to the analyzed vehicle, none of the fuels fulfilled the requirements. One ought to emphasize that the said norm was related to the NEDC test procedure, characterized with lower accuracy and, thus, the vehicle was able to fulfil the emission limits on the basis thereof. In connection with the perspective of lowering admissible emission limits for newly manufactured vehicles by 2030 (reduction of the limit by over 40% as compared with the one applicable in 2015), implementation of low emission solutions in all possible aspects will be necessary.
- According to the LCA analysis, FAME turned out to be the fuel with the lowest total emission. Yet, its production has the greatest environmental impact. It should be noted that production of long chain fatty acid esters from oil plants is competitive towards the food related purpose of such plants. The second place was taken by biobutanol, whose total carbon dioxide emission was nearly by one half lower than that of classical diesel fuel. Diesel fuel demonstrated the highest values of emitted CO_2 among all analyzed fuels, including the stage of production process.
- As fuel based on lignocellulose, biobutanol appears a promising energy medium, whose advantage comes from lower carbon dioxide emission as compared with conventional fuel which, in the light of stringent requirements and high declared CO_2 reduction levels, speaks very much to its advantage and encourages deeper theoretical and practical research into commercial application. Moreover, production of biobutanol from waste biomass carries additional possibilities to utilize useless matter.
- The physical and chemical properties of the biofuel and its percentage share in the fuel mixture have a significant impact on the course of the combustion process, self-ignition reactions and the rate of heat release, and consequently on gas emissions.
- The developed simulation constitutes a useful tool for initial research or planning of real experiments. It may be an element of a more comprehensive system or an independent system.
- Basing on the presented results it can also be concluded that decisions taken with respect to the processes of fuel production management should include the biobutanol produced from lignocellulosic biomass as an effective additive to the fuel or even as a fuel itself assuring positive environmental impact.
- The use of this type of fuel brings also the social effect since the biomass wastes are used for fuel production instead of edible parts of the agricultural crops.
- All the conclusions mentioned above indicate lignocellulose origin as promising for production of biofuel mitigating the carbon dioxide emission as well as avoiding consumption of edible parts of plants as raw material for biofuel production. This should be accepted as an indication for technology management as well as political decisions.

The available literature did not contain an approach that would link the actual research carried out on the engine for extensive changes in operating parameters. Other authors only presented the results of some tests but did not refer to the WLTP type approval tests, which are the condition for the approval of a given vehicle for use on public roads. The article cites the results of such studies, e.g., [12,13,58,59]. The present paper was aimed towards obtaining the answer whether, based on the operational characteristics of the engine, it is possible to estimate the behavior of the vehicle under the conditions of the WLTP dynamic test. The correctness of the simulation results was verified by reviewing the literature and referring to permissible emission limits. Currently, the authors are preparing a stand to perform full tests as part of driving tests.

Energies **2020**, *13*, 561

Author Contributions: Conceptualization, K.T., R.M., and O.O.; Methodology, O.O., K.T., and R.M.; Validation, A.W. and A.Ś.; Investigation, A.W. and A.Ś.; Writing—original draft preparation, K.T., O.O., and K.B.; Funding acquisition, A.Ś. All authors have read and agreed to the published version of the manuscript.

Funding: The authors wish to express gratitude to Lublin University of Technology for financial support given to the present publication (Antoni Świć). The research was carried out under financial support obtained from the research subsidy of the Faculty of Engineering Management (WIZ) of Bialystok University of Technology. From the grant no. WZ/WIZ/4/2019 (Olga Orynycz, Andrzej Wasiak).

Conflicts of Interest: The authors declare no conflict of interest. The funders had no role in the design of the study; in the collection, analyses, or interpretation of data; in the writing of the manuscript; or in the decision to publish the results.

References

1. Ferreira, J.A.; Brancoli, P.; Agnihotri, S.; Bolton, K.; Taherzadeh, M.J. A review of integration strategies of lignocelluloses and other wastes in 1st generation bioethanol processes. *Process. Biochem.* **2018**, *75*, 173–186. [CrossRef]

2. Kapanji, K.K.; Haigh, K.F.; Görgens, J.F. Techno-economic analysis of chemically catalysed lignocellulose biorefineries at a typical sugar mill: Sorbitol or glucaric acid and electricity co-production. *Bioresour. Technol.* **2019**, *289*, 121635. [CrossRef] [PubMed]

3. Tucki, K.; Orynycz, O.; Świć, A.; Mitoraj-Wojtanek, M. The Development of Electromobility in Poland and EU States as a Tool for Management of CO_2 Emissions. *Energies* **2019**, *12*, 2942. [CrossRef]

4. Chiaramonti, D.; Goumas, T. Impacts on industrial-scale market deployment of advanced biofuels and recycled carbon fuels from the EU Renewable Energy Directive II. *Appl. Energy* **2019**, *251*, 113351. [CrossRef]

5. Vanhala, P.; Bergström, I.; Haaspuro, T.; Kortelainen, P.; Holmberg, M.; Forsius, M. Boreal forests can have a remarkable role in reducing greenhouse gas emissions locally: Land use-related and anthropogenic greenhouse gas emissions and sinks at the municipal level. *Sci. Total Environ.* **2016**, *557*, 51–57. [CrossRef]

6. Venturi, S.; Tassi, F.; Cabassi, J.; Gioli, B.; Baronti, S.; Vaselli, O.; Caponi, C.; Vagnoli, C.; Picchi, G.; Zaldei, A.; et al. Seasonal and diurnal variations of greenhouse gases in Florence (Italy): Inferring sources and sinks from carbon isotopic ratios. *Sci. Total Environ.* **2020**, *698*, 134245. [CrossRef]

7. Krzywonos, M.; Tucki, K.; Wojdalski, J.; Kupczyk, A.; Sikora, M. Analysis of Properties of Synthetic Hydrocarbons Produced Using the ETG Method and Selected Conventional Biofuels Made in Poland in the Context of Environmental Effects Achieved. *Rocz. Ochr. Środowiska* **2017**, *19*, 394–410.

8. Schleussner, C.F.; Rogelj, J.; Schaeffer, M.; Lissner, T.; Licker, R.; Fischer, E.M.; Knutti, R.; Levermann, A.; Frieler, K.; Hare, W. Science and policy characteristics of the Paris Agreement temperature goal. *Nat. Clim. Chang.* **2016**, *6*, 827–835. [CrossRef]

9. FCCC/CP/2015/L.9/Rev.1. Adoption of the Paris Agreement. Available online: https://undocs.org/ (accessed on 16 November 2019).

10. Zak, A.; Golisz, E.; Tucki, K.; Borowski, P. Perspectives of biofuel sector development in Poland in comparision to CO_2 emission standards. *J. Agribus. Rural Dev.* **2014**, *3*, 299–312.

11. Valente, A.; Iribarren, D.; Candelaresi, D.; Spazzafumo, G.; Dufour, J. Using harmonised life-cycle indicators to explore the role of hydrogen in the environmental performance of fuel cell electric vehicles. *Int. J. Hydrogen Energy* **2019**. [CrossRef]

12. Chłopek, Z.; Biedrzycki, J.; Lasocki, J.; Wójcik, P. Pollutant emissions from combustion engine of motor vehicle tested in driving cycles simulating real–world driving conditions. *Zesz. Nauk. Inst. Pojazdów Politech. Warsz.* **2013**, *92*, 67–76.

13. Chłopek, Z.; Biedrzycki, J.; Lasocki, J.; Wójcik, P. The correlative studies of the pollutant emission and fuel consumption in type-approval tests. *TTS Tech. Transp. Szyn.* **2015**, *22*, 268–271.

14. Kupczyk, A.; Mączyńska, J.; Redlarski, G.; Tucki, K.; Bączyk, A.; Rutkowski, D. Selected Aspects of Biofuels Market and the Electromobility Development in Poland: Current Trends and Forecasting Changes. *Appl. Sci.* **2019**, *9*, 254. [CrossRef]

15. Tucki, K.; Mruk, R.; Orynycz, O.; Gola, A. The Effects of Pressure and Temperature on the Process of Auto-Ignition and Combustion of Rape Oil and Its Mixtures. *Sustainability* **2019**, *11*, 3451. [CrossRef]

16. PE/48/2018/REV/1 Directive (EU) 2018/2001 of the European Parliament and of the Council of 11 December 2018 on the Promotion of the Use of Energy from Renewable Sources (Text with EEA Relevance). Available online: https://eur-lex.europa.eu/ (accessed on 16 November 2019).
17. Górski, K.; Olszewski, W.; Lotko, W. Alcohols and ethers as fuels for diesel engines. *Czas. Tech. Mech.* **2008**, *105*, 13–24.
18. Bannikov, M.; Gollani, S.E.; Vasilen, I. Effect of alcohol additives on diesel engine performance and emissions. *Mater. Methods Technol.* **2015**, *9*, 8–19.
19. Tucki, K.; Mruk, R.; Orynycz, O.; Wasiak, A.; Botwinska, K.; Gola, A. Simulation of the Operation of a Spark Ignition Engine Fueled with Various Biofuels and Its Contribution to Technology Management. *Sustainability* **2019**, *11*, 2799. [CrossRef]
20. Rahman, Q.M.; Zhang, B.; Wang, L.; Shahbazi, A. A combined pretreatment, fermentation and ethanol-assisted liquefaction process for production of biofuel from *Chlorella* sp. *Fuel* **2019**, *257*, 116026. [CrossRef]
21. Mączyńska, J.; Krzywonos, M.; Kupczyk, A.; Tucki, K.; Sikora, M.; Pińkowska, H.; Bączyk, A.; Wielewska, I. Production and use of biofuels for transport in Poland and Brazil—The case of bioethanol. *Fuel* **2019**, *241*, 989–996. [CrossRef]
22. Lian, X.; Li, Y.; Zhu, J.; Zou, Y.; An, D.; Wang, Q. Fabrication of Au-decorated SnO2 nanoparticles with enhanced n-buthanol gas sensing properties. *Mater. Sci. Semicond. Process.* **2019**, *101*, 198–205. [CrossRef]
23. Ayad, S.M.; Belchior, C.R.; Da Silva, G.L.; Lucena, R.S.; Carreira, E.S.; De Miranda, P.E. Analysis of performance parameters of an ethanol fueled spark ignition engine operating with hydrogen enrichment. *Int. J. Hydrogen Energy* **2019**. [CrossRef]
24. Lapuerta, M.; Adrover, J.J.H.; Fernández-Rodríguez, D.; Cova-Bonillo, A. Autoignition of blends of n-butanol and ethanol with diesel or biodiesel fuels in a constant-volume combustion chamber. *Energy* **2017**, *118*, 613–621. [CrossRef]
25. Lapuerta, M.; Rodríguez-Fernández, J.; Fernández-Rodríguez, D.; Patiño-Camino, R. Modeling viscosity of butanol and ethanol blends with diesel and biodiesel fuels. *Fuel* **2017**, *199*, 332–338. [CrossRef]
26. Kamiński, W.; Tomczak, E.; Górak, A. Biobutanol—Production and purification methods. *Ecol. Chem. Eng. S* **2011**, *18*, 31–37.
27. Ezeji, T.C.; Qureshi, N.; Blaschek, H.P. Bioproduction of butanol from biomass: From genes to bioreactors. *Curr. Opin. Biotechnol.* **2007**, *18*, 220–227. [CrossRef] [PubMed]
28. Patakova, P.; Maxa, D.; Rychtera, M.; Linhova, M.; Fribert, P.; Muzikova, Z.; Lipovsky, J.; Paulova, L.; Pospisil, M.; Sebor, G.; et al. Perspectives of Biobutanol Production and Use. Available online: https://www.intechopen.com/books/biofuel-s-engineering-process-technology/perspectives-of-biobutanol-production-and-use (accessed on 18 January 2020).
29. Ramey, D.E.; Yang, S.T. Production of Butyric Acid and Butanol from Biomass. Final Report Number DOE-ER86106. Work Performed Under: Contract No.: DE-F-G02-00ER86106. Available online: https://www.osti.gov/biblio/843183-production-butyric-acid-butanol-from-biomass (accessed on 16 November 2019).
30. Bringué, R.; Ramírez, E.; Iborra, M.; Tejero, J.; Cunill, F. Esterification of furfuryl alcohol to butyl levulinate over ion-exchange resins. *Fuel* **2019**, *257*, 116010. [CrossRef]
31. Orchillés, A.V.; Vercher, E.; Miguel, P.J.; González-Alfaro, V.; Llopis, F.J. Isobaric vapor-liquid equilibria for the extractive distillation of tert-butyl alcohol + water mixtures using 1-ethyl-3-methylimidazolium dicyanamide ionic liquid. *J. Chem. Thermodyn.* **2019**, *139*, 105866. [CrossRef]
32. Liu, H.; Wang, G.; Zhang, J. The Promising Fuel-Biobutanol. In *Liquid, Gaseous and Solid Biofuels—Conversion Techniques*, 1st ed.; Fang, Z., Ed.; IntechOpen: London, UK, 2013; Available online: https://www.intechopen.com/books/liquid-gaseous-and-solid-biofuels-conversion-techniques/the-promising-fuel-biobutanol (accessed on 16 November 2019). [CrossRef]
33. Mack, J.H.; Schüler, D.; Butt, R.H.; Dibble, R.W. Experimental investigation of butanol isomer combustion in Homogeneous Charge Compression Ignition (HCCI) engines. *Appl. Energy* **2016**, *165*, 612–626. [CrossRef]
34. N-Butanol—Safety Data Sheet. Available online: https://www.perstorp.com/~|}/media/files/perstorp/msds/n-butanol/msds_n-butanol_pol-6694.ashx (accessed on 16 November 2019).
35. Xiao, H.; Guo, F.; Li, S.; Wang, R.; Yang, X. Combustion performance and emission characteristics of a diesel engine burning biodiesel blended with n-butanol. *Fuel* **2019**, *258*, 115887. [CrossRef]
36. Hönig, V.; Kotek, M.; Mařík, J. Use of butanol as a fuel for internal combustion engines. *Agron. Res.* **2014**, *12*, 333–340.

37. Liang, X.; Zhong, A.; Sun, Z.; Han, D. Autoignition of n-heptane and butanol isomers blends in a constant volume combustion chamber. *Fuel* **2019**, *254*, 115638. [CrossRef]
38. Smerkowska, B. Biobutanol—Production and application in diesel engines. *Chemik* **2011**, *65*, 549–556.
39. Huzir, N.M.; Aziz, M.A.; Ismail, S.; Abdullah, B.; Mahmood, N.A.N.; Umor, N.; Muhammad, S.A.F.S. Agro-industrial waste to biobutanol production: Eco-friendly biofuels for next generation. *Renew. Sustain. Energy Rev.* **2018**, *94*, 476–485. [CrossRef]
40. Dürre, P. Biobutanol: An attractive biofuel. *Biotechnol. J. Healthc. Nutr. Technol.* **2007**, *2*, 1525–1534. [CrossRef]
41. Qureshi, N.; Ezeji, T.C. Butanol, 'a superior biofuel' production from agricultural residues (renewable biomass): Recent progress in technology. *Biofuels Bioprod. Biorefin.* **2008**, *2*, 319–330. [CrossRef]
42. Van Der Wal, H.; Sperber, B.L.; Houweling-Tan, B.; Bakker, R.R.; Brandenburg, W.; López-Contreras, A.M. Production of acetone, butanol, and ethanol from biomass of the green seaweed Ulva lactuca. *Bioresour. Technol.* **2013**, *128*, 431–437. [CrossRef]
43. Figueroa-Torres, G.M.; Mahmood, W.M.A.W.; Pittman, J.K.; Theodoropoulos, C. Microalgal biomass as a biorefinery platform for biobutanol and biodiesel production. *Biochem. Eng. J.* **2020**, *153*, 107396. [CrossRef]
44. California Biobutanol Multimedia Evaluation. Tier I. Report. Available online: http://www.arb.ca.gov/fuels/multimedia/020910biobutanoltierI.pdf (accessed on 28 December 2019).
45. Jin, C.; Yao, M.; Liu, H.; Lee, C.F.F.; Ji, J. Progress in the production and application of n-butanol as a biofuel. *Renew. Sustain. Energy Rev.* **2011**, *15*, 4080–4106. [CrossRef]
46. Baustian, J.; Wolf, L. Cold-Start/Warm-Up Vehicle Performance and Driveability Index for Gasolines Containing Isobutanol. *SAE Int. J. Fuels Lubr.* **2012**, *5*, 1300–1309. [CrossRef]
47. Pałuchowska, M. Biobutanol produced from biomass. *Nafta-Gaz* **2015**, *7*, 502–509.
48. Karavalakis, G.; Short, D.; Vu, D.; Russell, R.L.; Asa-Awuku, A.; Jung, H.; Johnson, K.C.; Durbin, T.D. The impact of ethanol and iso-butanol blends on gaseous and particulate emissions from two passenger cars equipped with spray-guided and wall-guided direct injection SI (spark ignition) engines. *Energy* **2015**, *82*, 168–179. [CrossRef]
49. Ugwoha, E.; Andrésen, J.M. Sorption and phase distribution of ethanol and butanol blended gasoline vapours in the vadose zone after release. *J. Environ. Sci.* **2014**, *26*, 608–616. [CrossRef]
50. Zhen, X.; Wang, Y.; Liu, D. Bio-butanol as a new generation of clean alternative fuel for SI (spark ignition) and CI (compression ignition) engines. *Renew. Energy* **2020**, *147*, 2494–2521. [CrossRef]
51. Karavalakis, G.; Short, D.; Vu, D.; Villela, M.; Asa-Awuku, A.; Durbin, T.D. Evaluating the regulated emissions, air toxics, ultrafine particles, and black carbon from SI-PFI and SI-DI vehicles operating on different ethanol and iso-butanol blends. *Fuel* **2014**, *128*, 410–421. [CrossRef]
52. Yun, H.; Choi, K.; Lee, C.S. Effects of biobutanol and biobutanol–diesel blends on combustion and emission characteristics in a passenger car diesel engine with pilot injection strategies. *Energy Convers. Manag.* **2016**, *111*, 79–88. [CrossRef]
53. Nayyar, A.; Sharma, D.; Soni, S.L.; Mathur, A. Experimental investigation of performance and emissions of a VCR diesel engine fuelled with n-butanol diesel blends under varying engine parameters. *Environ. Sci. Pollut. Res.* **2017**, *24*, 20315–20329. [CrossRef]
54. Swamy, R.L.; Chandrashekar, T.K.; Banapurmath, N.R.; Khandal, S.V. Impact of Diesel-butanol Blends on Performance and Emission of Diesel Engine. *Oil Gas Res.* **2015**, *1*, 101. Available online: https://www.omicsonline.org/open-access/impact-of-dieselbutanol-blends-on-performance-and-emission-of-dieselengine-ogr-1000101.php?aid=63149 (accessed on 28 December 2019).
55. Lapuerta, M.; Hernández, J.J.; Rodríguez-Fernández, J.; Barba, J.; Ramos, A.; Fernández-Rodríguez, D. Emission benefits from the use of n-butanol blends in a Euro 6 diesel engine. *Int. J. Engine Res.* **2017**, *19*, 1099–1112. [CrossRef]
56. Liu, H.; Wang, X.; Zhang, D.; Dong, F.; Liu, X.; Yang, Y.; Huang, H.; Wang, Y.; Wang, Q.; Zheng, Z. Investigation on Blending Effects of Gasoline Fuel with N-Butanol, DMF, and Ethanol on the Fuel Consumption and Harmful Emissions in a GDI Vehicle. *Energies* **2019**, *12*, 1845. [CrossRef]
57. Douvartzides, S.L.; Charisiou, N.D.; Papageridis, K.N.; Goula, M.A. Green Diesel: Biomass Feedstocks, Production Technologies, Catalytic Research, Fuel Properties and Performance in Compression Ignition Internal Combustion Engines. *Energies* **2019**, *12*, 809. [CrossRef]

58. Verma, P.; Stevanovic, S.; Zare, A.; Dwivedi, G.; Van, T.C.; Davidson, M.; Rainey, T.; Brown, R.J.; Ristovski, Z.D. An Overview of the Influence of Biodiesel, Alcohols, and Various Oxygenated Additives on the Particulate Matter Emissions from Diesel Engines. *Energies* **2019**, *12*, 1987. [CrossRef]
59. Elfasakhany, A.; Mahrous, A.F. Performance and emissions assessment of n-butanol–methanol–gasoline blends as a fuel in spark-ignition engines. *Alex. Eng. J.* **2016**, *55*, 3015–3024. [CrossRef]
60. Black, G.; Curran, H.; Pichon, S.; Simmie, J.; Zhukov, V. Bio-butanol: Combustion properties and detailed chemical kinetic model. *Combust. Flame* **2010**, *157*, 363–373. [CrossRef]
61. Frassoldati, A.; Grana, R.; Faravelli, T.; Ranzi, E.; Oßwald, P.; Kohse-Höinghaus, K.; Osswald, P. Detailed kinetic modeling of the combustion of the four butanol isomers in premixed low-pressure flames. *Combust. Flame* **2012**, *159*, 2295–2311. [CrossRef]
62. Pexa, M.; Čedík, J.; Hönig, V.; Pražan, R. Lignocellulosic Biobutanol as Fuel for Diesel Engines. *BioResources* **2016**, *11*, 6006–6016. [CrossRef]
63. Čedík, J.; Pexa, M.; Mařík, J.; Hönig, V.; Horníčková, Š.; Kubín, K. Influence of butanol and FAME blends on operational characteristics of compression ignition engine. *Agron. Res.* **2015**, *13*, 541–549.
64. Peterka, B.; Pexa, M.; Čedík, J.; Mader, D.; Kotek, M. Comparison of exhaust emissions and fuel consumption of small combustion engine of portable generator operated on petrol and biobutanol. *Agron. Res.* **2017**, *15*, 1162–1169.
65. Zhang, Y.; Huang, R.; Xu, S.; Huang, Y.; Huang, S.; Ma, Y.; Wang, Z. The effect of different n-butanol-fatty acid methyl esters (FAME) blends on puffing characteristics. *Fuel* **2017**, *208*, 30–40. [CrossRef]
66. Hönig, V.; Pexa, M.; Linhart, Z. Biobutanol Standardizing Biodiesel from Waste Animal Fat. *Pol. J. Environ. Stud.* **2015**, *24*, 2433–2439. [CrossRef]
67. Rakopoulos, D.C.; Rakopoulos, C.D.; Giakoumis, E.G.; Papagiannakis, R.G.; Kyritsis, D.C. Influence of properties of various common bio-fuels on the combustion and emission characteristics of high-speed DI (direct injection) diesel engine: Vegetable oil, bio-diesel, ethanol, n-butanol, diethyl ether. *Energy* **2014**, *73*, 354–366. [CrossRef]
68. Rakopoulos, D.; Rakopoulos, C.; Hountalas, D.; Kakaras, E.; Giakoumis, E.; Papagiannakis, R. Investigation of the performance and emissions of bus engine operating on butanol/diesel fuel blends. *Fuel* **2010**, *89*, 2781–2790. [CrossRef]
69. Tracking Clean Energy Progress. Available online: https://www.iea.org/tcep/transport/biofuels/ (accessed on 16 November 2019).
70. COM. 112—A Roadmap for Moving to a Competitive Low Carbon Economy in 2050. 2011. Available online: https://www.eea.europa.eu/policy-documents/com-2011-112-a-roadmap (accessed on 16 November 2019).
71. Galadima, A.; Muraza, O. Zeolite catalyst design for the conversion of glucose to furans and other renewable fuels. *Fuel* **2019**, *258*, 115851. [CrossRef]
72. Statistical Yearbook of Forestry. Available online: https://stat.gov.pl (accessed on 16 November 2019).
73. National Forestry Accounting Plan (NFAP). Developed by the Team for the Elaboration of National Plans Related to Accounting for Greenhouse Gas Emissions and Removals Resulting from Forestry Activities. Warsaw, 2018. Available online: https://bip.mos.gov.pl (accessed on 16 November 2019).
74. Kożuch, A.; Banaś, J.; Zięba, S.; Bujoczek, L. Changes in the synthetic index of sustainable forest management at the level of regional directorates of the State Forests in 1993–2013. *For. Res. Pap.* **2018**, *79*, 229–236. [CrossRef]
75. Forests in Poland. 2017. Available online: http://www.lasy.gov.pl/pl/informacje/publikacje/in-english/forests-in-poland/lasy-w-polsce-2017-en.pdf (accessed on 16 November 2019).
76. Państwowe Gospodarstwo Leśne Lasy Państwowe. Forest Report in Poland; 2018. Available online: https://bip.lasy.gov.pl/pl/bip/px_~{}raport_o_lasach_2018_do_bip.pdf (accessed on 16 November 2019).
77. Państwowe Gospodarstwo Leśne Lasy Państwowe. Forest Report in Poland; 2010. Available online: https://www.bdl.lasy.gov.pl/portal/Media/Default/Publikacje/raport_o_stanie_lasow_2010.pdf (accessed on 16 November 2019).
78. Głowacki, S.; Bazylik, W.; Sojak, M. Urban Green as a Source of Biomass for Energy Purposes. *Ciepłownictwo Ogrzew. Went.* **2013**, *44*, 206–209.
79. European Commission. Environment. Eco-Management and Audit Scheme (EMAS). Available online: https://ec.europa.eu/environment/emas/index_en.htm (accessed on 2 January 2020).

80. Environmental Management Systems. The ISO 14001 Standard. Available online: https://www.pcbc.gov.pl/pl/uslugi/certyfikacja-systemow-zarzadzania/pn-en-iso-14001 (accessed on 2 January 2020).
81. Brito, M.; Martins, F. Life cycle assessment of butanol production. *Fuel* **2017**, *208*, 476–482. [CrossRef]
82. Levasseur, A.; Bahn, O.; Beloin-Saint-Pierre, D.; Marinova, M.; Vaillancourt, K. Assessing butanol from integrated forest biorefinery: A combined techno-economic and life cycle approach. *Appl. Energy* **2017**, *198*, 440–452. [CrossRef]
83. Brandão, M.; Martin, M.; Cowie, A.; Hamelin, L.; Zamagni, A. Consequential Life Cycle Assessment: What, How, and Why? In *Encyclopedia of Sustainable Technologies*; Abraham, M.A., Ed.; Elsevier: Amsterdam, The Netherlands, 2017; pp. 277–284.
84. Brander, M. Comparative analysis of attributional corporate greenhouse gas accounting, consequential life cycle assessment, and project/policy level accounting: A bioenergy case study. *J. Clean. Prod.* **2017**, *167*, 1401–1414. [CrossRef]
85. Environmental Priority Strategies (EPS). Available online: http://www.gabi-software.com/international/support/gabi/gabi-lcia-documentation/environmental-priority-strategies-eps/ (accessed on 2 January 2020).
86. CML-IA Characterisation Factors. Available online: https://www.universiteitleiden.nl/onderzoek/onderzoeksoutput/wiskunde-en-natuurwetenschappen/cml-cml-ia-characterisation-factors (accessed on 2 January 2020).
87. Eco-Indicator 99 Method. Available online: https://www.sciencedirect.com/topics/engineering/eco-indicator (accessed on 2 January 2020).
88. Jolliet, O.; Margni, M.; Charles, R.; Humbert, S.; Payet, J.; Rebitzer, G.; Rosenbaum, R. IMPACT 2002+: A new life cycle impact assessment methodology. *Int. J. Life Cycle Assess.* **2003**, *8*, 324–330. [CrossRef]
89. ReCiPe. Available online: https://www.pre-sustainability.com/recipe (accessed on 2 January 2020).
90. Wiesen, K.; Saurat, M.; Lettenmeier, M. Calculating the Material Input per Service Unit using the Ecoinvent database. *Int. J. Perform. Eng.* **2014**, *10*, 357–366.
91. LCA. Software for Fact-Based Sustainability. Available online: https://simapro.com/ (accessed on 2 January 2020).
92. GaBi Software. Available online: http://www.gabi-software.com/ce-eu-english/software/gabi-software/ (accessed on 2 January 2020).
93. Material Flow Analysis & Life Cycle Assessment with the Software Umberto. Available online: https://www.ifu.com/en/umberto/ (accessed on 2 January 2020).
94. Biernat, K. Perspectives for global development of biofuel technologies to 2050. *Chemik* **2012**, *66*, 1178–1189.
95. Piekarski, W.; Zając, G. Possibility of the use of liquid biofuels as a supply to the internal combustion engines. *Autobusy Tech. Eksploat. Syst. Transp.* **2011**, *12*, 347–354.
96. Biernat, K.; Jeziorkowski, A. Problems in supplying modern internal combustion engine with biofuels. *Studia Ecol. Bioethicae* **2008**, *6*, 307–329.
97. Fiat Panda Misc Documents Accessories Brochure PDF. Available online: https://manuals.co/ (accessed on 16 November 2019).
98. Ambrozik, A.; Kurczyński, D.; Łagowski, P.; Warianek, M. The toxicity of combustion gas from the Fiat 1.3 Multijet engine operating following the load characteristics and fed with rape oil esters. *Proc. Inst. Veh.* **2016**, *1*, 23–36.
99. Ambrozik, A.; Ambrozik, T.; Kurczyński, D. Load characteristics of turbocharged 1.3 Multijet engine. *Postępy Nauk. Tech.* **2012**, *15*, 7–20.
100. Merkisz, J.; Pielecha, I.; Pielecha, J.; Brudnicki, K. On-Road Exhaust Emissions from Passenger Cars Fitted with a Start-Stop System. *Arch. Transp.* **2011**, *23*, 37–46. [CrossRef]
101. Zhu, T.; Wu, Y.; Li, B.; Zong, C.; Li, J. Simulation Research on the Start-stop System of Hybrid Electric Vehicle. *J. Adv. Veh. Eng.* **2017**, *3*, 55–64.
102. Regulation (EC) No 443/2009 of the European Parliament and of the Council of 23 April 2009 Setting Emission Performance Standards for New Passenger Cars as Part of the Community's Integrated Approach to Reduce CO_2 Emissions from Light-Duty Vehicles (Text with EEA Relevance). Available online: https://eur-lex.europa.eu/ (accessed on 16 November 2019).
103. Proposal for a Regulation of the European Parliament and of the Council Setting Emission Performance Standards for New Passenger Cars and for New Light Commercial Vehicles as Part of the Union's Integrated Approach to Reduce CO_2 Emissions from Light-Duty Vehicles and Amending Regulation (EC) No 715/2007 (Recast). Available online: https://eur-lex.europa.eu/ (accessed on 16 November 2019).

104. CO$_2$ Targets Are Becoming Ever More Demanding Worldwide. Available online: https://www.daimler.com/sustainability/vehicles/climate-protection/wltp/wltp-part-5.html (accessed on 16 November 2019).
105. Panda Catalog. Available online: http://www.auto-alex.pl/files/23.pdf (accessed on 16 November 2019).
106. Baczewski, K.; Kałdoński, T. *Paliwa do Silników o Zapłonie Samoczynnym*, 2nd ed.; Wydawnictwa Komunikacji i Łączności: Warszawa, Poland, 2017; pp. 50–210.
107. Gwardiak, H.; Rozycki, K.; Ruszkarska, M.; Tylus, J.; Walisiewicz-Niedbalska, W. Evaluation of fatty acid methyl esters (FAME) obtained from various feedstock. *Oilseed Crops* **2011**, *32*, 137–147.
108. Tucki, K.; Mruk, R.; Orynycz, O.; Wasiak, A.; Swić, A. Thermodynamic Fundamentals for Fuel Production Management. *Sustainability* **2019**, *11*, 4449. [CrossRef]
109. International Energy Agency (IEA). Advanced Motor Fuels. Butanol, Properties. Available online: https://www.iea-amf.org/content/fuel_information/butanol/properties (accessed on 16 November 2019).
110. Kruczyński, S.W.; Orliński, P.; Orliński, S. The effects of powering the agricultural engine with mixture diesel fuels with biobutanol to economic and energy signs of its work. *Logistka* **2011**, *6*, 1–8.
111. Campbell, S.L.; Chancelier, J.P.; Nikoukhah, R. *Modeling and Simulation in Scilab/Scicos with ScicosLab 4.4*, 1st ed.; Springer: New York, NY, USA, 2006; pp. 73–155.
112. Lachowicz, C.T. *Matlab, Scilab, Maxima: Opis i Przykłady Zastosowań*, 1st ed.; Oficyna Wydawnicza Politechniki Opolskiej: Opole, Poland, 2005; pp. 120–310.
113. Brozi, A. *Scilab w Przykładach*, 1st ed.; Wydawnictwo Nakom: Poznań, Poland, 2007; pp. 62–209.
114. Pavlovic, J.; Marotta, A.; Ciuffo, B. CO$_2$ emissions and energy demands of vehicles tested under the NEDC and the new WLTP type approval test procedures. *Appl. Energy* **2016**, *177*, 661–670. [CrossRef]
115. Pavlovic, J.; Ciuffo, B.; Fontaras, G.; Valverde, V.; Marotta, A. How much difference in type-approval CO$_2$ emissions from passenger cars in Europe can be expected from changing to the new test procedure (NEDC vs. WLTP)? *Transp. Res. Part A Policy Pract.* **2018**, *111*, 136–147. [CrossRef]
116. Merkisz, J. Real Road Tests—Exhaust Emission Results from Passenger Cars. *J. KONES Powertrain Transp.* **2011**, *18*, 253–260.
117. Barlow, T.; Latham, S.; McCrae, I.; Boulter, P. A Reference Book of Driving Cycles for Use in the Measurement of Road Vehicle Emissions. Available online: https://trl.co.uk/reports/PPR354 (accessed on 16 November 2019).
118. Regulation No 83 of the Economic Commission for Europe of the United Nations (UN/ECE)—Uniform Provisions Concerning the Approval of Vehicles with Regard to the Emission of Pollutants According to Engine Fuel Requirements. Available online: https://eur-lex.europa.eu/ (accessed on 16 November 2019).
119. Tokgoz, S.; Laborde, D. Indirect Land Use Change Debate: What Did We Learn? *Curr. Sustain. Energy Rep.* **2014**, *1*, 104–110. [CrossRef]
120. Khanna, M.; Zilberman, D. Modeling the land-use and greenhouse-gas implications of biofuels. *Clim. Chang. Econ.* **2012**, *3*, 1250016. [CrossRef]
121. Plevin, R.J. Assessing the Climate Effects of Biofuels Using Integrated Assessment Models, Part I: Methodological Considerations. *J. Ind. Ecol.* **2016**, *21*, 1478–1487. [CrossRef]
122. Styles, D.; Gibbons, J.; Williams, A.P.; Dauber, J.; Stichnothe, H.; Urban, B.; Chadwick, D.R.; Jones, D.L. Consequential life cycle assessment of biogas, biofuel and biomass energy options within an arable crop rotation. *GCB Bioenergy* **2015**, *7*, 1305–1320. [CrossRef]
123. Decicco, J.M. Methodological Issues Regarding Biofuels and Carbon Uptake. *Sustainability* **2018**, *10*, 1581. [CrossRef]
124. De Kleine, R.D.; Anderson, J.E.; Kim, H.C.; Wallington, T.J. Life cycle assessment is the most relevant framework to evaluate biofuel greenhouse gas burdens. *Biofuels Bioprod. Biorefin.* **2017**, *11*, 407–416. [CrossRef]
125. Glensor, K.; Muñoz, B.; Rosa, M. Life-Cycle Assessment of Brazilian Transport Biofuel and Electrification Pathways. *Sustainability* **2019**, *11*, 6332. [CrossRef]
126. Van Der Voet, E.; Lifset, R.J.; Luo, L. Life-cycle assessment of biofuels, convergence and divergence. *Biofuels* **2010**, *1*, 435–449. [CrossRef]
127. Jones, C.; Gilbert, P.; Raugei, M.; Mander, S.; Leccisi, E. An approach to prospective consequential life cycle assessment and net energy analysis of distributed electricity generation. *Energy Policy* **2017**, *100*, 350–358. [CrossRef]

128. Menten, F.; Chèze, B.; Patouillard, L.; Bouvart, F. A review of LCA greenhouse gas emissions results for advanced biofuels: The use of meta-regression analysis. *Renew. Sustain. Energy Rev.* **2013**, *26*, 108–134. [CrossRef]

129. Guyon, O. Methodology for the Life Cycle Assessment of a Car-sharing Service. Available online: https://www.diva-portal.org/smash/get/diva2:1183366/FULLTEXT01.pdf (accessed on 16 November 2019).

130. ISO. Environmental Management—Life cycle assessment—Requirements and Guidelines. Available online: https://www.iso.org/standard/38498.html (accessed on 16 November 2019).

131. Volkswagen AG Group Research. The Caddy—Environmental Commendation Background Report. Available online: https://www.a-pointduurzaamheid.nl (accessed on 16 November 2019).

132. Volkswagen AG Group Research. The Golf—Environmental Commendation Background Report. Available online: https://www.a-pointduurzaamheid.nl/files/4914/2348/7329/e_golf_Env_Comm.pdf (accessed on 16 November 2019).

133. Ambrozik, A.; Kurczyński, D. Load characteristics in ad3.152 ur engine fuelled with mineral and biogenous blends. *J. KONES Powertrain Transp.* **2006**, *13*, 183–194.

134. Labeckas, G.; Slavinskas, S. The effect of diesel fuel blending with rapeseed oil and rme on engine performance and exhaust emissions. *J. KONES Powertrain Transp.* **2005**, *12*, 187–194.

Article

Geographical Potential of Solar Thermochemical Jet Fuel Production

Christoph Falter *, Niklas Scharfenberg and Antoine Habersetzer

Bauhaus Luftfahrt e.V., Willy-Messerschmitt-Str. 1, 82024 Taufkirchen, Germany;
niklas.scharfenberg@bauhaus-luftfahrt.net (N.S.); antoine.habersetzer@bauhaus-luftfahrt.net (A.H.)
* Correspondence: christoph.falter@bauhaus-luftfahrt.net

Received: 2 December 2019; Accepted: 3 February 2020; Published: 12 February 2020

Abstract: The solar thermochemical fuel pathway offers the possibility to defossilize the transportation sector by producing renewable fuels that emit significantly less greenhouse gases than conventional fuels over the whole life cycle. Especially for the aviation sector, the availability of renewable liquid hydrocarbon fuels enables climate impact goals to be reached. In this paper, both the geographical potential and life-cycle fuel production costs are analyzed. The assessment of the geographical potential of solar thermochemical fuels excludes areas based on sustainability criteria such as competing land use, protected areas, slope, or shifting sands. On the remaining suitable areas, the production potential surpasses the current global jet fuel demand by a factor of more than fifty, enabling all but one country to cover its own demand. In many cases, a single country can even supply the world demand for jet fuel. A dedicated economic model expresses the life-cycle fuel production costs as a function of the location, taking into account local financial conditions by estimating the national costs of capital. It is found that the lowest production costs are to be expected in Israel, Chile, Spain, and the USA, through a combination of high solar irradiation and low-level capital costs. The thermochemical energy conversion efficiency also has a strong influence on the costs, scaling the size of the solar concentrator. Increasing the efficiency from 15% to 25%, the production costs are reduced by about 20%. In the baseline case, the global jet fuel demand could be covered at costs between 1.58 and 1.83 €/L with production locations in South America, the United States, and the Mediterranean region. The flat progression of the cost-supply curves indicates that production costs remain relatively constant even at very high production volumes.

Keywords: GIS; concentrated solar power; solar thermochemistry; life-cycle costs; cost supply; geographical potential; sustainable; alternative

---

1. Introduction

A goal set for this century is the transition of the transportation sector from a fossil energy base to a renewable one. This goal is mainly motivated by the necessity to limit climate change through a reduction of carbon dioxide emissions and by the limited long-term supply security of fossil fuels. This transition is very challenging to achieve as today by far the largest share of the energy used in this sector is provided by fossil fuels [1] and a switch to a radically different technology will necessarily involve large investments into infrastructure and propulsion technology [2]. Nevertheless, electro-mobility is projected to drastically increase its share in the ground-based transportation sector [1], which would enable one to mainly use electricity for the propulsion of light-duty vehicles. In heavy-duty and airborne transportation, liquid hydrocarbon fuels are an ideal energy carrier due to their high energy density and favorable handling properties, as well as the existing global supply infrastructure. Especially for aviation, a transition to hydrogen or batteries is not as easy to implement as for cars because of the much stricter requirements for low weight and the volume of the energy carrier. It is therefore desirable to produce a liquid hydrocarbon fuel from renewable primary energy that can be used with the current infrastructure and propulsion technology. Among the different options, the use of solar energy is promising due to its widespread availability and already existing economic conversion technologies into heat and electricity. In recent years, electrochemical and thermochemical pathways have shown interesting results. Here, the focus is on the latter due to its high energy conversion potential [3,4] and significant experimental progress [3,5–7].

As solar energy is in principle able to cover the global energy demand, its conversion into liquid fuels could also easily cover the fuel demand of global aviation. As the production of solar thermochemical fuels requires only sunlight, water, and carbon dioxide, it could give a range of countries the possibility to produce their own environmentally-friendly fuels without having to rely on imports from oil-producing countries. However, as sunlight is unevenly distributed over the surface of the earth, there are regions that are more suitable for solar fuel production than others. It is therefore interesting to analyze the dependency of the fuel production costs on geographical location and to quantify the production potential in different regions of the earth.

In the literature, the geographical production potential of solar electricity with concentrated solar power (CSP) has been analyzed e.g., in [8–14] and that of PV electricity in [15]. The power-to-liquid (PtL) pathway uses water electrolysis to produce hydrogen and converts it with CO_2 to liquid fuels using reverse water gas shift and the Fischer–Tropsch process and is therefore technically related to the solar thermochemical pathway. The potential and cost of PtL fuels have been determined in [16–19] and it is found that the pathway is in principle scalable to meet the largest demands at costs of about 3.2 €/L today or 1.4 €/L assuming low-cost renewable electricity in 2050 [19]. To the best of the authors' knowledge, the geographical potential in combination with the cost of solar thermochemical fuels have not been published so far, as only cost estimates exist for specific single locations. Thus, the focus is on the geographical variability of solar thermochemical fuel production characteristics. The theoretical amount of fuels that can be produced along with the estimated production costs for the most interesting regions of the earth is presented. To this end, a geographic information system (GIS)-based approach is used in combination with technical and economic models of the fuel production process. Studies have been performed on the geographical potential of solar technologies for electricity production [8,20,21], and on the potential of biogenic sustainable energy [22–25], but none on the particular case of solar thermochemical fuels. After the exclusion of unsuitable areas due to sustainability criteria, the linkage between production potential and production cost is shown on a regional and national level with cost-supply curves. With the information shown in this work, the regional and global potential for solar fuel production is quantified and the life-cycle costs for single regions and countries can be estimated, establishing a useful tool for decision-makers in the area of alternative fuels.

The Solar Thermochemical Pathway

In the solar thermochemical pathway, solar energy is concentrated to provide heat to a redox cycle of a metal oxide operating between temperatures of about 1000 K and 1800 K. Cerium oxide is chosen as the reactive material, which is placed in a solar cavity reactor. Through the input of solar heat, the reactor is cyclically heated to the upper process temperature, where the material releases a part of its stored oxygen under a reduced oxygen partial pressure in the reactor. This is achieved through the removal of oxygen with a vacuum pump. In the second step of the cycle, the temperature of the reactor is reduced, and water and carbon dioxide are introduced, which are subsequently split into oxygen and synthesis gas (hydrogen and carbon monoxide). Cerium oxide returns to its initial state by absorbing the evolving oxygen, and the synthesis gas can then be converted into liquid fuels by the Fischer–Tropsch process. A schematic of the solar thermochemical fuel production process is shown in Figure 1.

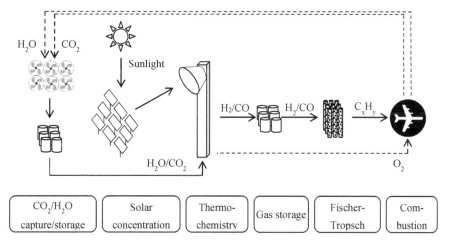

Figure 1. Schematic of the solar thermochemical fuel production process. H_2O and CO_2 are captured from the air and provided to the solar thermochemical conversion, where synthesis gas (H_2 + CO) is produced. The synthesis gas is then turned into liquid fuels via the Fischer–Tropsch process.

The redox reactions are shown in Equations (1)–(3):

$$\frac{1}{\delta_{red} - \delta_{ox}} CeO_{2-\delta_{ox}} \rightarrow \frac{1}{\delta_{red} - \delta_{ox}} CeO_{2-\delta_{red}} + \frac{1}{2}O_2 \tag{1}$$

$$\frac{1}{\delta_{red} - \delta_{ox}} CeO_{2-\delta_{red}} + H_2O \rightarrow \frac{1}{\delta_{red} - \delta_{ox}} CeO_{2-\delta_{ox}} + H_2 \tag{2}$$

$$\frac{1}{\delta_{red} - \delta_{ox}} CeO_{2-\delta_{red}} + CO_2 \rightarrow \frac{1}{\delta_{red} - \delta_{ox}} CeO_{2-\delta_{ox}} + CO$$

$$\begin{aligned} H_2O &\rightarrow \tfrac{1}{2}O_2 + H_2 \\ CO_2 &\rightarrow \tfrac{1}{2}O_2 + CO \end{aligned} \tag{3}$$

Water and carbon dioxide can in principle be provided from any source, whereas direct air capture is an attractive option for future implementations due to its environmental performance and the avoidance of long-range gas transport. Through chemical adsorption to a sorbent, both water and carbon dioxide are captured, whereas even in dry regions, the amount of water captured surpasses the amount of captured carbon dioxide [26]. The technology is currently in a demonstration phase with early commercial applications e.g., by Climeworks [27].

Solar thermochemical fuel production has been the subject of ongoing research and has experienced a significant increase in reactor efficiency from a value of 0.8% [5] in 2008 to 1.7% [6] in 2012, and most recently to above 5% [7] (at a potential exceeding 50% [28]). These results were achieved through an improved material and reactor design, enabling higher heat and mass transfer rates in the reactor. Especially the introduction of different length scales for the porosity of the reactive material has reduced the time scales for heating up the material, which reduces the required power input while maintaining a high surface area needed for quick reoxidation [3,6,7]. Further development of the technology is likely to comprise material morphology and the introduction of an effective heat exchanger to reduce the energy input to the cycle. Heat recuperation from the solid phase has been shown to be vital for the achievement of higher efficiencies due to the comparable small oxygen nonstoichiometry of non-volatile material cycles [4,29–32]. Several promising reactor designs exist that could further enhance the conversion efficiency by the incorporation of solid-solid heat exchange [31,33,34] and by using particles for higher flexibility of the design [35,36]. The material properties are also sought to be improved through doping with specific elements [28,37–39] or through completely new material combinations [40–42]. To achieve an economic production process, it is assumed that efficiency of roughly 20% has to be reached for the thermochemical conversion [43,44], whereas the exact value is subject to detailed economic modeling and also depends on economic and political framework conditions. While experimental values are still significantly below this threshold, theoretical analyses have shown that values even beyond 20% are possible [45].

2. Methodology

In this section, the methodology chosen for the determination of suitable areas and production costs of solar thermochemical fuels is described. In general, the analysis was limited to regions with a high level of direct normal irradiation (DNI), i.e., the USA, the Andes region in South America (including Chile, Bolivia, Argentina, and Peru), the Mediterranean region (MED; including Southern Europe, the Middle East, and North Africa), Southern Africa (including South Africa, Botswana, Angola, and Namibia), and Australia. First, a list of exclusion criteria was defined for the determination of the available areas. Then the remaining net areas were combined with the available DNI and a conversion factor to determine the production potential. In a second step, an economic model was applied to express the production costs of solar thermochemical fuels as a function of the location. In the following, these steps are discussed in more detail.

2.1. Determination of Suitable Areas for Solar Thermochemical Fuel Production

The software QGIS [46] was used as a GIS-based tool for the net area calculations. Similar to the calculations performed in the MED CSP study [8] for the determination of suitable areas for CSP plants in the Mediterranean, the following areas were excluded: areas with existing ground structures, water bodies, shifting sands, slopes ≥5%, protected areas, as well as areas covered by forest, closed shrubland, woody savannas, wetland, cropland, urban settlements, or snow/ice. Consequently, allowed land types are open shrubland, savannas, and barren or sparsely vegetated land that do not fall under the restrictions listed above. Regarding land cover, data from the University of Arizona were used with a resolution of 15 arc seconds that assigns each pixel a type of land coverage based on the highest confidence for the years 2001–2010 [47]. Vector data on protected areas are taken from the World Database on Protected Areas [48], a joint project of the United Nations Environment Program and the International Union for the Conservation of Nature. Protected areas comprise national protected areas recognized by the government, areas designated under conventions, privately protected areas, and community-conserved territories. The data on shifting sand dunes and quicksand is taken from the Food and Agriculture Organization of the United Nations (FAO) [49]. The vector graphics for country borders are retrieved from the Database of Global Administrative Areas (GADM) [50] and the coastlines are from the Global Self-consistent, Hierarchical, High-resolution Geography Database [51]. The slope data was derived from a digital elevation model of the National Oceanic and Atmospheric Administration (NOAA) [52].

A map with the net available areas is created by subtracting all excluded areas described above. On the remaining land, the direct normal irradiation (DNI) is mapped with data from the Global Solar Atlas [53] with a resolution of 1 km^2. The data cover latitudes between 60° N and 45° S, whereas the incline of the satellite images prevents an accurate assessment of cloud covers outside of this area. Local DNI values are derived from a combination of a clear sky model that calculates the local solar energy flux based on the position of the sun, altitude, concentration of aerosols, water vapor content, and ozone, and satellite images that detect the cloud cover. The primary grid resolution is 3–7 km, which is downscaled to a resolution of 1 km. The satellite data represents the average DNI over a range of years, being 1999–2015 for the USA and South America, 1994–2015 for Europe and Africa, 1999–2015 for the eastern part of the MED region, and 2007–2015 for Australia.

2.2. Determination of Life-Cycle Production Costs

To determine the production costs of one liter of solar thermochemical jet fuel, an economic model was developed based on the levelized cost of energy (LCOE) [54]. The base year of the analysis is 2017.

Due to the long lifetime of the plant of 25 years, the time value of money has to be taken into account. This is achieved through the annualization of the investment and operational costs, which requires the definition of an interest rate. This interest rate is equivalent to the weighted average cost of capital (WACC) and is comprised of an equity rate and a debt rate. Consequently, the calculation of an annual value of the total costs of the plant is the sum of the annualized investment costs and the yearly operation and maintenance costs. The total annual costs are then divided by the annual production volume of fuel. Specifically, the equations used are the following:

$$\text{LCOE} = \frac{I + PV_{\text{O\&M}}}{Q \times A} \tag{4}$$

I denotes the investment costs assumed to be paid at the beginning of the plant lifetime, $PV_{\text{O\&M}}$ is the present value of the operational costs of the plant, which accumulate over the lifetime, Q is the annual amount of fuel produced, and $A = (1 - (1 + i)^{-n})\, i^{-1}$ is the annuity factor. i denotes the interest rate and n the lifetime of the plant:

$$A = \frac{1 - (1+i)^{-n}}{i} \tag{5}$$

The investment costs are assumed to be financed 60% through debt and 40% through equity, which is equivalent to a WACC of $0.6\,d + 0.4e$, with the interest rate for debt d and for equity e. The cost of debt and equity can vary substantially between countries due to differences regarding political, budgetary, and macroeconomic stability, as well as financial market efficiency [55]. However, it is not possible to determine the exact cost of debt and equity for a specific investment project, as this is determined by financial market actors. Rather, a meaningful comparison of country-specific cost of debt and equity needs to rely on suitable proxy indicators. Consequently, the weighted average costs of capital determined here should be understood as estimates. In the following, it is described which indicators, data sources, and methods are used.

The equity interest rate is assumed to be the sum of the government bond yields and an equity risk premium (ERP). Essentially, the ERP is an indication for "the compensation investors require to make them indifferent at the margin between holding the risky market portfolio and a risk-free bond" [56]. More specifically, one can assume that government bonds represent the benchmark for risk-free bonds. Thus, to estimate equity interest rates, one can sum up country-specific government bond yields with country-specific equity risk premiums. Government bond yield data are retrieved from different financial analyst organizations [57,58]. For the latter, data compiled by Damodaran [59] are used, whereas ERP values are calculated from mature market premiums, which are adapted by country risk premiums. Even though ERP values are hard to determine [60], the authors think that the dataset used here is adequate for the performance of a general comparison of the cost of capital across different countries. However, it is necessary to emphasize that diverse models and sources for ERP exist and that ERP estimates can vary substantially between sources [56,60]. For an in-depth discussion of the costs of capital for specific projects, more detailed information would be needed.

The debt interest rate is taken to be the nominal bank lending rate. As the primary dataset, the International Monetary Fund's (IMF) International Financial Statistics [61] are used. However, as this dataset is not complete, lending rates for missing countries are determined from the economic analysis platform CEIC [62]. To further increase the robustness of the variable, a second dataset for bank lending rates is retrieved from the CIA world factbook [63]. The final values for lending rates by country were determined by taking the average of both datasets. Inflation, retrieved from the IMF [64] was averaged over five years (2013–2017) to level out short-term effects. The estimates for interest rates and the resulting weighted average cost of capital is shown in Table 1 for selected countries, whereas the information for all countries is given in the annex.

Note that, for some countries and indicators, data were missing. As multiple imputation methods prove to be a useful method to adequately deal with missing data (especially in comparison to simpler methods, like dropping observations with missing values), multiple imputation using chained equations is employed here [65]. For an adequate estimation of the missing values of interest [65] the main variables are used, that is, interest rate, inflation rate, equity risk premiums, and government bond yields, as well as auxiliary predictors, namely GDP per capita [66], and several macroeconomic indicators from the World Economic Forum Global Competitiveness Indicators database [55]. Thirty imputed datasets are created, from which the final imputed values were calculated.

Table 1. Overview of estimated interest rates, inflation, and the weighted average cost of capital for selected countries (2017).

Countries	Debt Interest Rate [%]	Government Bond Yields [%]	Equity Risk Premium [%]	Inflation (5-Year Average) [%]	Nominal Weighted Average Cost of Capital [%]
USA	4.0	3.1	5.1	1.3	5.7
Spain	2.2	1.6	7.3	0.5	4.9
Chile	4.6	5.0	5.8	3.3	7.1
Israel	3.4	2.2	5.9	0.2	5.3
Egypt	18.8	17.5	12.6	12.3	23.3
South Africa	10.4	9.5	7.6	5.6	13.1

In the literature, not many studies vary the financing costs across countries, even though the impact on the LCOE is large [67]. For comparison, WACC values for solar projects in the literature are 7% (nominal) in the US SunShot Vision Study [68,69], and 7.5% for OECD countries and China, and 10% (real) for other countries in a study by the IRENA [70]. Labordena et al. used default values for the real equity rate of return recommended by the UNFCCC to calculate country-specific WACC [67] for CSP projects in sub-Saharan Africa.

The data in Table 1 show that the costs of capital for a fuel production plant substantially vary across different countries. This is because some regions provide a more stable environment for investments than others, which is expressed in the interest rate estimates and inflation. High costs of capital represent the risk involved for the investor and thus increase the total costs of the investment. If interest rates and inflation are high, the resulting production costs of jet fuel will be high as well. This may render a very sunny location unfeasible for solar fuel production because of the resulting high costs of capital. An example is Egypt, which has the sunniest locations in the MED region but has very high interest rate estimates, which increase the production costs over those of Israel and even Spain, which have a smaller solar resource. This is also illustrated with an estimation of the weighted average costs of capital for the analyzed countries and the global DNI in Figure 2.

It is important to note that these estimated WACC were only a snapshot for the year 2017 and rely on aggregated proxy indicators. Thus, the cost of capital for a specific project in the future may be quite different from the estimates presented here. Our aim here is not to give a precise prediction of future investment environments, but to sensitize for the importance of country-specific variations in the cost of capital. The sensitivity of the production costs on the inflation rate and weighted average costs of capital is shown in Appendix A.5.

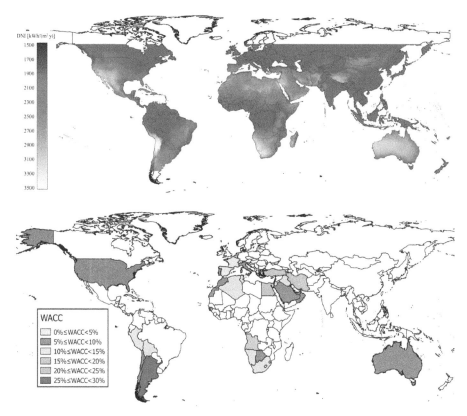

Figure 2. Direct normal irradiation (DNI, **top**) and estimates of nominal weighted average costs of capital (WACC, **bottom**) in the selected countries for the analysis. The most favorable locations for solar thermochemical fuel production are those with a high DNI and low WACC.

2.2.1. Investment Costs and O&M Costs

The investment and operating costs were estimated based on a model developed in [44] for a plant producing 1050 barrels per day (bpd) of jet fuel with naphtha as a by-product (see Appendix A.4. for a list of assumptions and costs). Naphtha is assumed to be sold at a relative price of 0.8 with respect to the production costs of jet fuel, corresponding to the correlation of market prices seen for conventional fuels [71,72]. The model is updated to include jet vacuum pumps instead of mechanical pumps and reforming of the light hydrocarbons produced in the FT reactor is performed. The use of vacuum pumps enabled significantly higher experimental efficiencies [7], where the costs for the pumps for this study are taken from [73]. The reforming of the light hydrocarbons enables a higher yield from the produced syngas. Furthermore, the thermochemical reactors and the gas-to-liquid conversion steps were modeled in more detail than in the previous publication using dedicated Matlab and Aspen models. CO_2 and H_2O are captured from the air based on chemical adsorption to an amine-functionalized solid sorbent [74] and are stored in tanks before being supplied to the thermochemical reactor operating at an efficiency of 19.0% (ratio of higher heating value of gases leaving the reactor to concentrated solar energy entering including auxiliary energy for heating the reacting gases) excluding the energy for vacuum pumping and gas separation. The specific cost of the air capture unit was assumed to be 350 €/t at fixed operational costs of 40 €/t, resulting in the long-term target costs of about 100 €/t [74]. The thermochemical reactors were assumed here to have a unit size of 50 kW at a cost of 14 €/kW each for the reactor shell excluding ceria, which does not preclude larger unit sizes. This cost estimate assumes a scaling exponent of 0.6 when scaling up from a single 50 kW-unit to the 2.8 GW of total thermal power of the current plant design, which corresponds to about 56,000 units of 50-kW reactors. The reactors complete 16 redox cycles per day using ceria as the reactive material, which is replaced after 500 completed cycles. This cyclability has been demonstrated in tests on a small material sample [75], while close to 300 cycles were achieved with a 4-kW reactor [6]. The efficiency of the reactors was assumed to be 19% in the baseline, which has not been achieved so far but is a realistic target for the development in the near to medium-term development. Taking into account the tower structure, the reactor shell and the reactive material ceria, the total cost of the receivers and tower is estimated to be 65 €/kW of thermal input power. This result assumes that the reactors can be mass-produced, achieving low production costs. The heat required in the reactor is supplied via the concentration of solar energy from a heliostat field with an aperture area of 8.8 million m^2 and the electricity with a dedicated CSP tower plant with a heliostat area of 0.57 million m^2. The size of the heliostat fields is dependent on the local DNI and thus changes with the location of the plant. A unit cost of 0.04 €/kWh$_{el}$ was assumed for solar electricity, which is somewhat lower than the best-projected values of below 0.10 €/kWh$_{el}$ from the newest CSP plants in China [76] and above the best values for PV electricity of 0.02 €/kWh$_{el}$ [77,78] and could represent a future combined price of PV and CSP. The heliostat costs are set to 100 €/m^2, which is a likely cost value for the considered time frame of about 10–20 years in the future. Hydrogen and carbon monoxide coming from the reactor were stored and supplied to the Fischer–Tropsch (FT) plant, which has specific investment costs of 23,000 €/bpd and O&M costs of 4 € per barrel [79]. Light hydrocarbons from the FT unit are steam reformed and fed back into the FT process. The finished products were transported to the sea with a pipeline, which is assumed to be built for this purpose at specific costs of 90 €/m [80,81]. It is assumed that a direct connection to the sea from the plant location can be made and altitude differences in between are neglected. Final transportation of the products by ship over a distance of 1000 km is also taken into account at a unit cost of 0.8 cents per liter of liquid product [82].

For a plant of 1050 bpd, the jet fuel capacity located in a sunny region with a DNI of 2500 kWh/(m^2 y) and at a distance of 250 km to the sea, the total investment costs are 1.50 billion €. Of this sum, 59% are for the heliostat field, 8% for the thermochemical reactors, 7% for the CO$_2$ capture units and syngas storage, respectively, and other smaller contributions. The operational costs sum up to 55.2 million € per year, whereas the largest cost contributors are the heliostat field (32%), the capture of H$_2$O and CO$_2$ (21%), the generation of electricity (21%), and syngas storage (6%).

For plant operation and maintenance, a workforce is required with workers, technicians, engineers, management, and administration. To determine the number of direct jobs created, estimates of the total number of jobs in CSP plants are taken as a reference: the International Renewable Energy Agency (IRENA) estimates that 0.6–1.33 jobs are created per MW of the electrical output power of a CSP plant [83]. Sooriyaarachchi et al. analyze job creation in different renewable energy technologies and estimate 0.3 jobs per MWp [84]. In another publication of the IRENA, 0.2 jobs created per GWh are mentioned [85], while a report by Applied Analysis gives a number of 500–720 jobs per 2 GW plant [86]. The Energy Sector Management and Assistance Program (ESMAP) indicates 13–20 jobs created for a 500,000 m^2 solar trough power plant [87]. Finally, Solar Reserve published numbers regarding their newly built CSP plant in Port Augusta, Australia, mentioning 50 jobs for the plant of 135 MW [88]. These estimates, converted to the equivalent plant size of 2.8 GW thermal power (or 1.1 GW electrical power), give a corridor of about 200–500 jobs created for the baseline case plant with one larger estimate up to 1700 jobs. Using these figures as guidance, a total workforce of 396 is assumed, distributed into 246 workers, 100 technicians, 30 engineers, 15 clerks, and 5 managers. With data from the International Labor Organization (ILO) on annual salaries by occupation and country [89], the cost of labor is derived (see annex).

2.2.2. Dependency of Production Costs on DNI, Distance to Sea, and Reactor Efficiency

The production costs depend to a significant degree on the heliostat field, the size of which is determined by the DNI at the plant site and the reactor efficiency. At higher irradiation and a constant output of the plant, a smaller heliostat field can provide the required amount of concentrated energy to the solar reactors and thus the investment and O&M costs are decreased. At a less favorable site, the heliostat field may need to be enlarged to deliver the specified power, which will increase the production costs. The influence of reactor efficiency is like scaling the heliostat field and the associated costs. As a second geographical variable, the distance to the sea has an impact on the production costs because it determines the length of the pipeline for the distribution of the products. It is assumed that a direct connection to the sea can be made by the pipeline, neglecting altitude differences. At the sea, the products are then transported to their final location by ship over a distance of 1000 km. The dependency of production costs on DNI, distance from the sea and reactor efficiency is shown in Figure 3, whereas all other parameters are left constant.

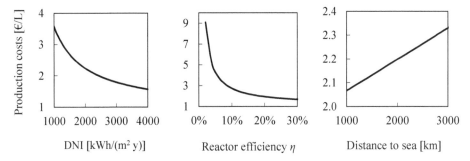

Figure 3. Production costs as a function of solar irradiation (DNI, **left**), reactor efficiency (η, **center**), and distance to sea (**right**), whereas the latter determines the length of the product pipeline.

From a DNI of 2000 kWh/(m^2 y), which was a value found in the South of Spain, to the best locations in the Andes at 3500 kWh/(m^2 y), the production costs dropped by about 25%. Equally, increasing reactor efficiency from 10% to 25%, the costs dropped by 35%. From a location directly at the sea to one that is 2000 km away from the shore, the production costs rose by about 15%. As a distance of 2000 km from the sea is found rarely, the DNI was very likely to have a stronger influence on the production costs. From the dependency of the production costs on these variables, a first conclusion can be drawn: an efficient reactor at a location with high solar irradiation located at or near the sea will reduce production costs under otherwise constant conditions.

3. Results

Using the methodology described above, the suitable areas in the regions of the Mediterranean region (MED), the USA, South America, South Africa, and Australia are identified. Having identified the suitable areas, the production volumes and production costs are determined.

3.1. Suitable Areas for Solar Thermochemical Fuel Production

In the following maps, the exclusion of areas is indicated with colors (green: land use, red: protected areas, yellow: shifting sand dunes, black: slope ≥5%), while the suitable areas within national boundaries are shown in white.

The excluded and suitable areas in the MED region are shown in Figure 4. In Europe, only very little area is available due to other land uses such as agriculture and livestock farming, and due to large areas being protected from other uses. Suitable areas are mostly found in Spain and Turkey. In northern Africa and the Arabian Peninsula, there are vast suitable desert areas where the climate is too dry for agricultural use. Excluded areas are found in Morocco and throughout the other countries in northern Africa and the Middle East mostly due to shifting sands and protected areas. Agricultural areas are found mostly along the shore in Morocco and Algeria, along the river Nile, and on the eastern Mediterranean coast.

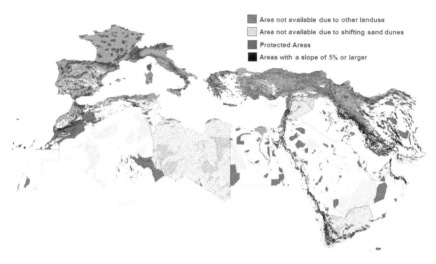

Figure 4. Suitable areas for solar thermochemical fuel production in the MED region. Excluded areas comprise other land use (**green**), protected areas (**red**), ≥5% slope (**black**), and shifting sands (**yellow**). Libya, Syria, and Yemen are excluded from the analysis due to missing socio-economic data.

In the USA (Figure 5), most of the land is being used for agriculture and livestock farming or is protected in national parks. In the Appalachian and Rocky Mountain regions, the terrain may be too steep for the construction of a fuel production plant. The suitable areas are found in the southwest, where the solar irradiation is also highest. While most of the country is unavailable for solar thermochemical fuel production, the remaining suitable areas are nevertheless considerable in size and could enable large fuel production. In South America (Figure 5), the most favorable locations are found in the Andes region due to the particularly high DNI in the desert at high altitudes. Due to the mountainous region, however, many locations have too steep slopes to be used for the construction of a plant. Large regions are also excluded due to site protection, agriculture, and livestock farming, leaving the largest suitable areas in eastern Argentina. In the eastern part of South Africa (Figure 6) mainly grassland and cropland prohibit the production of solar fuels. In the northern part of Angola, the land is covered mainly by woody savannas, evergreen broadleaf forest, and cropland at the coast. There are also a number of protected areas such as the Kruger national park or the Kalahari Desert, which are excluded, as well as shifting sands along the coast in Namibia and Angola. Nevertheless, there remain large connected areas in all four countries that are theoretically available. In Australia's (Figure 6), southeast, cropland and evergreen broadleaf forest, and protected areas all over the country significantly reduce the available areas, which are however still the largest found in a single country.

Figure 5. Suitable areas for solar thermochemical fuel production (**white**) in the USA (**left**) and Chile, Bolivia, Peru and Argentina (**right**). Excluded areas comprise other land use (**green**), protected areas (**red**), ≥5% slope (**black**), and shifting sands (**yellow**).

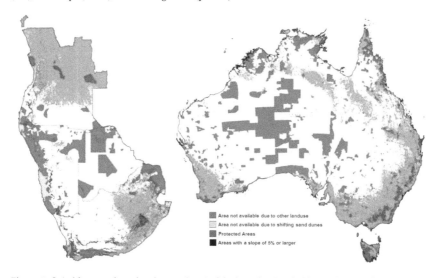

Figure 6. Suitable areas for solar thermochemical fuel production (**white**) in South Africa, Namibia, Angola and Botswana (**left**) and in Australia (**right**). Excluded areas comprise other land use (**green**), protected areas (**red**), ≥5% slope (**black**), and shifting sands (**yellow**).

3.2. Production Potential on Suitable Areas

After the identification of suitable areas, it is possible to derive the production potential of solar thermochemical fuels. For this purpose, a conversion factor is defined that expresses the fraction of DNI that is turned into jet fuel. This conversion factor is comprised of the energy conversion efficiency of the fuel production process and the land-use factor, whereas the latter is the area of the heliostats divided by the area of the land, and which is assumed to be 25% [90]. The energy conversion efficiency of direct solar irradiation to jet fuel is 2.46% (based on updated calculations of [44], see Appendix A.3 for more information), where naphtha is produced as a by-product. The conversion factor is then 0.25 × 0.0246 = 0.61%. Thus, at a common annual DNI of 2500 kWh/(m^2 y), 2500 kWh/(m^2 y) × 3.6 MJ/kWh

× 0.61%/33.4 MJ/L = 1.66 L of jet fuel are produced per square meter and year. For the calculations of the production potential, it is assumed that the conversion factor is constant and thus independent of the DNI. In Table 2 the sum of the areas suitable for fuel production is shown for different regions of the earth together with the production volumes of solar fuels.

Table 2. Calculated suitable areas and production potential of solar thermochemical fuels in different regions.

	Suitable Area for Solar Fuel Production [10^6 km^2]	Suitable Area/Calculated Country Area [%]	Production Volume on Suitable Areas [10^3 Mt/y] [b]	Production Volume/World Demand [-]
MED region	5.71	52	6.13	20.4
Australia	4.49	59	5.81	19.4
South Africa, Botswana, Namibia, Angola	1.87	48	2.38	7.9
Chile, Argentina, Peru, Bolivia [a]	1.24	23	1.47	4.9
USA [a]	0.599	8	0.777	2.6

[a] The deviation from the FAO area is largely due to the data being restricted to latitudes between 60° N and 45° S.
[b] Current global jet fuel consumption is about 300 Mt/y [91].

In the Appendix A (Table A1), the calculated total areas of the countries in the different regions are compared with the areas indicated by the FAO [92] to test the calculation methodology of the geographic information system. The largest suitable areas for solar fuel production are found in the MED region with close to 6 million km^2, where the largest areas ($\geq 10^6$ km^2) are found in North Africa and the Middle East, i.e., in Algeria, Western Sahara, and in Saudi Arabia. As a single country, Australia has by far the largest potential with about 4.5 million km^2 of suitable area corresponding to more than half of the country area. In the MED region, the share of suitable areas in the respective country areas is also high with >50% on average, whereas this value is mainly achieved in North Africa and the Middle East, where large regions are available in the desert. The value found for the countries in southern Africa is also close to 50%. In the USA, however, below 10% of the country is suitable for solar fuel production, which is due to large areas being used for agriculture and farming. The suitable and most interesting sites are found in California, Nevada, Arizona, New Mexico, and Texas, which is where the solar irradiation is highest. For comparison, the SunShot study found suitable areas of about 0.226 × 10^6 km^2 in the Southwest of the US with more stringent exclusion criteria for slope and a lower bound for the DNI [68], while another study found 0.985 × 10^6 km^2 [14]. The potential found in this study is in between these values, where the differences stem from the definition of the exclusion criteria. In the analyzed countries in South America, only about 20% of the total areas are found to be suitable, which is mainly due to high slopes in the Andes region, agricultural land use, pastureland, and protected areas.

The volume of fuel that can be produced on a given area can be derived by multiplying the area by the specific productivity, which is dependent on the local solar irradiation. In Table 2, the sum of the production volumes that could be achieved on the suitable areas in the chosen regions is shown. The largest volumes can be produced in the MED region due to the largest available areas and quite high DNI values in the Sahara Desert and on the Arabian Peninsula, whereas the highest DNI is found on the Sinai Peninsula in Egypt. The production volumes reach 6.1 × 10^3 Mt/y, which is more than 20 times the current value of global jet fuel consumption (about 300 Mt/y [91]). The other regions achieve smaller production volumes of about 5.8 × 10^3 Mt/y in Australia, 1.5 × 10^3 Mt/y in southern America, 2.4 × 10^6 Mt/y in southern Africa, and 0.8 × 10^3 Mt/y in the USA, which is still more than twice the global jet fuel consumption. Therefore, all areas and even single countries are in principle capable of producing enough jet fuel to cover the current global demand, which means that almost all countries (except France) are able to cover their own demand to become independent of oil imports for aviation and possibly also for other transport sectors. With this huge potential for fuel production, it is possible to limit the choice of areas to the best locations with the lowest production cost.

3.3. Life-Cycle Production Costs of Solar Thermochemical Jet Fuel

In the following, the production costs of solar thermochemical jet fuel are analyzed in the most interesting regions. The results are displayed as cost-supply curves, i.e., production costs per liter are shown as a function of the respective amount of fuel that can be produced at that cost. For production costs maps of the regions, please refer to Appendix A.2. To facilitate the analysis of the production volumes, the current jet fuel consumption (2015–2017, depending on data availability [93], green line in the graphs) in the selected regions is given and also the global jet fuel consumption [91] of 300 Mt/y (red line in the graphs). As the fuel production potential exceeds the world demand in all of the analyzed regions, the graphs are limited to supply up to the world demand and the region of the graphs beyond the world demand are shown only in the Appendix A.1. for completeness.

In Figure 7, cost-supply curves for the selected regions are shown for reactor efficiencies of 0.15, 0.19, and 0.25. In general, the change in efficiency leads to a shift of the cost-supply curves, where a higher efficiency reduces the costs due to the smaller heliostat field required. Among the interesting process parameters, reactor efficiency has been chosen to be varied here due to its large influence on costs and its large potential improvement over the current state of the art. In the following, the results are discussed only for the case of a reactor efficiency of 0.19. In the MED region (Southern Europe, northern Africa and the Middle East), there is a large production potential for solar fuels, which easily exceeds the global jet fuel demand (see Appendix A.1 for full curves). The curve indicates that the local jet fuel demand could be covered with production costs between 1.58 and 1.82 €/L, with average costs of 1.75 €/L, and the global demand at costs of 1.58–2.09 €/L (average 1.98 €/L). The lowest production costs are found in regions with high solar irradiation and favorable financial boundary conditions (interest rate estimates and inflation), i.e., in Israel and Spain, while the highest DNI occurs in Egypt and in Saudi Arabia. However, in the latter countries, the cost of capital is comparably high, which leads to overall higher production costs. It should be noted that the chosen economic model is based on the assumption of market conditions for the cost of capital and does not take into account other political boundary conditions such as purchase agreements or financial support.

The curve thus starts at the lowest costs in the most favorable production locations of Israel and Spain and further includes mainly Morocco, Western Sahara, and Saudi Arabia. The inclination of the cost-supply curve indicates the availability of areas on which fuel can be produced at the specific costs. A steep curve section thus represents a comparably restricted supply capability, which is due to the limited areas available at the respective irradiation and financial conditions. A flat curve section, on the other hand, indicates that large amounts of fuel can be produced at rather constant production costs.

The USA has quite highly irradiated areas in the southwest and very favorable costs of capital, enabling low-cost production of solar fuels. At the best locations, production costs of under 1.75 €/L can be achieved, whereas the national demand for jet fuel could be covered at costs between 1.74 and 1.82 €/L, and at average costs of 1.79 €/L (Figure 7b). If the global jet fuel demand were to be covered, the costs would be between 1.74 and 1.88 €/L at average costs of 1.84 €/L.

In Australia, very large areas with high solar irradiation are suitable for solar fuel production, which leads to the highest production potential of a single country. The national and global demand of jet fuel can therefore easily be covered with a fraction of the available areas at costs of 1.86 €/L for the former and between 1.86 and 1.91 €/L for the latter at an average of 1.90 €/L. The best locations are found in the northwest, where the highest values of solar irradiation are found and the distance to the sea is short. Somewhat higher costs of capital and labor costs in Australia prevent even lower production costs. If the financial boundary conditions could be further improved, the country could become a very favorable location for the production of solar fuels.

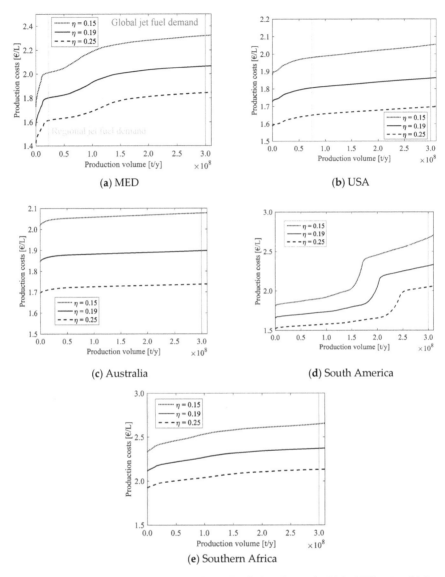

Figure 7. Cost-supply curves of solar thermochemical jet fuel production for (**a**) the MED region, (**b**) the USA, (**c**) Australia, (**d**) South America, and (**e**) Southern Africa. Three values of reactor efficiency are distinguished: 0.15, 0.19, 0.25. The current jet fuel demand is also shown in the selected countries [93] (**green line**) and world jet fuel demand [91] (**red line**). Note that the y-axes represent different intervals.

The South American Andes region is an excellent location for the conversion of solar energy with the highest values of direct solar irradiation found globally. From a technical point of view, this region is therefore ideally suited for solar fuel production with maximum DNI values exceeding 3500 kWh/(m² y). On the other hand, the regional financial preconditions are somewhat less favorable with relatively high estimates of interest rates: the weighted average cost of capital is between 7.1% in Chile and 29.3% in Argentina. The resulting production costs thus rise to higher values than in the best locations of the MED region with DNI values of about 2700 kWh/(m² y) in Israel with a WACC of 5.3%. Nevertheless, Chile offers exceptionally good conditions for the production of solar fuels, which enable production costs below 1.67 €/L. The regional demand of the selected countries can be covered at average costs of 1.68 €/L and the global demand at 1.67–2.34 €/L (average: 1.94 €/L). In the graph, the production locations are found only in Chile and Bolivia due to their favorable conditions. The production costs rise slowly from low values in Chile and experience a sharp rise when the least favorable production locations in Chile are selected. At an accumulated production volume of about 2.25×10^8 t/y, the production moves into Bolivia, which has higher production costs than Chile according to its higher cost of capital (nominal WACC: 11.8%). The slope of the production costs rises again slowly up to the production volume in Bolivia corresponding to the global demand.

In southern Africa, the solar irradiation is also relatively high, making this region likewise favorable from a technical point of view. The production potential is quite high, exceeding those for the selected countries in South America. However, the cost of capital is comparably high with the lowest nominal WACC found in Botswana at 8.8%, about 13% in South Africa and Namibia, and up to 18.4% in Angola. The resulting production costs start at 2.13 €/L in Botswana, whereas locations in Namibia and South Africa have the lowest costs of 2.51 €/L and 2.53 €/L, respectively. Angola has very high costs of capital, preventing the production of solar jet fuel below 4.8 €/L. In southern Africa, the regional demand could, therefore, be covered at average costs of 2.14 €/L and the global demand at 2.32 €/L using the best locations found in Botswana.

In general, the flat progression of the cost-supply curve suggests that large-scale production of solar fuels could be set in place at relatively constant production costs. For the sake of clarity and simplicity, we do not conceptualize a dynamic cost model, which incorporates decreasing production costs with increasing production capacities. Thus, more sophisticated models could incorporate economies of scale and learning effects to better capture cost dynamics with increasing capacities. Note that a precondition for its deployment on a GW-scale is the availability of all technologies at a high technology readiness level (TRL). While still at lower TRL today, the current development of CO_2 air capture and reactor technology is promising [7,27]. At a GW-scale, the availability of ceria could provide a challenge and require its replacement with other available and tested reactive materials such as perovskites [40].

4. Discussion

In the preceding analysis, the cost of capital was geographically varied by modeling the local interest rates for debt and equity and using country-specific inflation rates. This leads to differing WACC estimates in each state. Under otherwise constant conditions, the production costs of solar jet fuel will be higher in a country with high costs of capital (i.e., a high WACC) than in a country with a low WACC. A deviation from this behavior may occur for locations with differing geophysical and macroeconomic conditions, e.g., different levels of solar irradiation and thus sizes of heliostat fields, or for countries with deviating inflation rates. The results shown above, therefore, reflect a change in estimated investment costs, in WACC, and inflation rates across countries. Besides these varying costs of capital, it is interesting to look at the case of constant nominal WACC across countries to analyze the effect on the cheapest production locations of solar jet fuel. This represents the case where e.g., the state acts as a guarantor for the investment into the fuel plant and thus enables a low cost of capital. In the following the nominal WACC is—hypothetically—held constant across countries at a level of 4% or 10%, respectively. The inflation rates are, however, still country-specific and are much harder to influence by governments. This is due to the fact that inflation is influenced by a variety of macroeconomic dynamics, and that central banks—the institution with the strongest means to influence inflation—often act independently from governments. This results in constant nominal WACC values but varying real WACC values across the locations. Note that locations with high inflation rates then show lower production costs. This counterintuitive result can be explained by the guaranteed *nominal* interest rates, which result in lower *real* interest rates at higher inflation. In other words, if the state finances investment projects at a certain nominal cost of capital (e.g., in countries with relatively high-interest rates), it will have to refinance itself through government bonds to a real interest rate, which is likely higher than the guaranteed nominal cost of capital.

It should be noted that a fluctuation in exchange rates could lead to partial compensation of this development. In other words, countries with high inflation rates could see their currency depreciate over time, which reduces the prices of domestically produced goods on global markets. However, as exchange rates do not only depend on the inflation rates in different countries, a quantification of this effect is rather difficult and out of the scope of this paper. In most countries, the inflation rate is between 0% and 5%, while in some cases it can reach 10% and above (i.e., in Egypt, Angola, and Iran). This should be kept in mind when interpreting the following results. The resulting cost-supply curves for the baseline case (with varying nominal WACC) and fixed nominal WACC values are shown in Figure 8, where the production location is indicated by the line color and the regional share of total fuel volume for each case is indicated with a pie chart for better legibility.

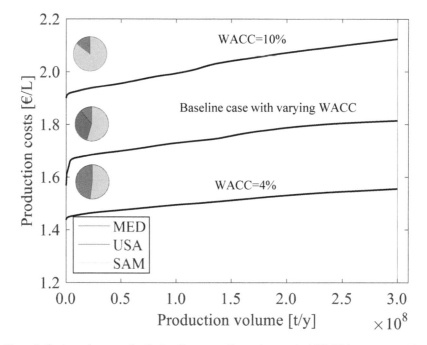

Figure 8. Cost-supply curves for the baseline case with varying nominal WACC between countries and constant nominal WACC in each country of 4% and 10%, respectively. The colors of the pie charts indicate the distribution of total production volumes over the regions for each case in the different regions of Europe, North Africa, and the Middle East (MED, **red**), the USA (USA, **blue**), and South America (SAM, **green**). The graph only shows production volumes up to the current global jet fuel demand (indicated by the vertical red line).

The baseline case starts with the lowest production costs in Israel and Chile, which both have comparably low nominal WACCs of about 5% and 7% and inflation of close to 0% and 3%, respectively (2013–2017). Despite not having the best solar resource in the MED region, Israel thus achieves very low production costs through its beneficially low costs of capital. Egypt, on the other hand, with the best solar resource in the region, has a much higher nominal WACC of 23.3% and an inflation rate of 12.3%, which results in prohibitively high production costs and therefore does not appear in the cost-supply curves shown above. (In the Appendix A.1, the cost-supply curves are shown for even larger production volumes, which then include all countries.) The baseline case further comprises locations in Spain (WACC$_{nom}$ = 4.9%, inflation rate = 0.5%) and the USA (WACC$_{nom}$ = 5.7%, inflation rate = 1.3%). The contributions of the single regions to the total production volume are shown in the pie chart in the graph and are about 55% for South America, 34% for the USA, and 11% for the MED region in the baseline case. The sharp increase in the curve at the low end of the production costs is due to the scarce availability of the best locations with very low costs of capital.

For constant nominal costs of capital of 4% and 10%, the cost-supply curves of solar jet fuel show, in general, a similar progression as for the baseline case but at different levels of production costs, which is due to the varying inflation rates, solar resources and distances to the sea across the countries. For WACC$_{nom}$ = 4%, the lowest production costs are below 1.45 €/L and found in Chile due to the superior solar irradiation, which leads to a smaller heliostat field and thus costs. For WACC$_{nom}$ = 10%, the best locations are found in Chile, which indicates that the higher Chilean inflation rate of 3.3% is compensated by the lower investment costs of the smaller heliostat field. At WACC$_{nom}$ = 4%, 48% of the fuel volume originates in the MED region and the remaining 52% in South America, mainly in Chile. At the higher level of capital costs, however, 85% of the volume is produced in South America and 15% in the MED region. This shift in production locations as a function of financial boundary conditions is explained in the following. If inflation rates and nominal WACC vary between countries (as in the baseline case), the locations with the lowest production costs of about 1.60 €/L are found in Israel, a country with a fairly high solar irradiation and very favorable financial conditions. The locations with the highest solar irradiation in Chile achieve larger production costs because of their higher costs of capital, which overcompensate the smaller investment and operational costs. If, on the other hand, the nominal WACC is fixed at 4% or 10%, Chile emerges as the country with the lowest production costs. With constant (nominal) interest rates across countries, the minimization of heliostat area in the sunniest areas leads to the most economical production location, while this becomes even more important towards higher interest rates. Inflation is still considered to be country-specific, giving an advantage to sunny countries with a stable environment for investments. The comparison of the two cases with fixed interest rates shows that at low costs of capital, the preferred locations are found equally in MED and SAM across most of the cost range, while at high costs of capital by far the most locations (especially the ones with the lowest costs) are in SAM and only very few in MED. The USA drop out of the list of preferred locations for fixed WACC due to other locations having even higher solar irradiation values and lower labor costs. A guarantee by the state for low costs of capital is therefore especially interesting in countries in the MED region that do not have the highest solar resource and also for countries with otherwise high costs of capital.

It should be kept in mind, however, that in this case the financial burden is shifted away from the plant owner and onto the institution that is enabling the attractive financial conditions. Nevertheless, especially in view of supply security and a potential deep reduction of greenhouse gas emissions from the aviation sector, this may prove to be an interesting option for facilitating the entry into the market of the technology.

The country-specific variation of costs of capital are only rough estimates for possible future market conditions. As a consequence, one should be rather careful with interpretations. The values can be used for a high-level comparison between countries, but they can change drastically in the future. The aim here is to sensitize the importance of country variations.

In addition, as the analysis focuses on the most promising countries for production, one might miss some interesting areas in other countries. For instance, smaller areas in Mexico, Western China, or Central Asia might hold promising production sites as well. This could further increase the production potential and lead to a more widespread diffusion of solar thermochemical jet fuel production. Future research could focus more specifically on the countries left out in this analysis.

5. Conclusions

A geographical assessment of solar thermochemical fuel production for the decarbonization of the transportation sector, especially aviation, is presented. A GIS-based methodology was developed to indicate the suitable areas for the production of solar fuels, whereas non-suitable areas were excluded based on criteria of existing land use such as agriculture or pastureland, slope, shifting sands, and protected areas. The remaining areas are found to have a huge production potential, which surpasses the world jet fuel demand by more than a factor of fifty. With an economic model, the life-cycle production costs are expressed as a function of the location, whereas local financial conditions are taken into account by estimating regional costs of capital. Cost-supply curves are then used to express the interplay between local production potentials and the respective production costs. The lowest production costs are found in Israel, Chile, and the USA. While not having the largest solar resource, the financial conditions are very favorable in the several countries of the Mediterranean region, with low costs of capital estimates and very low inflation rates. This partly compensates for the higher investment costs into a larger mirror field, which is required due to the smaller solar resource. A higher reactor efficiency has the same effect of reducing the solar field size, which in turn reduces investment costs. Increasing the efficiency from 15% to 25%, the production costs are reduced by about 20%. The baseline efficiency is assumed to be 19%, which has not been achieved so far but is a realistic target for the near-to medium-term technological development of solar reactors. In general, the cost-supply curves are relatively flat, which indicates that unlike for other fuel production pathways, which are limited by resource provision (i.e., biomass-based pathways), the solar thermochemical pathway can produce practically unlimited amounts of fuel at relatively constant costs. The work here presents for the first time a geographical analysis of possible production locations. It thus gives an indication of production costs, the best production locations, and the local, regional and global production potentials. With the presented information, it is possible to derive strategies for technological development and financial support for the production of solar thermochemical fuels on a regional and national level.

Author Contributions: C.F. conceived the research idea. N.S. and C.F. performed the calculations. A.H. conceived the adaptation of the financial model to different countries. All authors were involved in writing, editing and revising the manuscript. All authors have read and agreed to the published version of the manuscript.

Funding: The research leading to these results has received funding from the European Union's Horizon 2020 research and innovation program under grant agreement no. 654408.

Acknowledgments: The authors would like to acknowledge the support of Valentin Batteiger and Andreas Sizmann.

Conflicts of Interest: The authors declare no conflict of interest. The funders had no role in the design of the study; in the collection, analyses, or interpretation of data; in the writing of the manuscript, or in the decision to publish the results.

Appendix A

Table A1. Suitable areas and production potential for each country.

Region or country	Calculated Country Area/FAO Area [a] [%]	Suitable Area for Solar Fuel Production [km²]	Suitable Areas/Calculated Country Area [%]	Production Volume on Suitable Areas [t/y] [b]
MED region	97%	5,708,154	52%	6.13×10^9
Algeria	97%	1,634,220	71%	1.77×10^9
United Arab Emirates	98%	22,169	32%	2.17×10^7
Egypt	98%	623,294	64%	7.30×10^8
Western Sahara	100%	234,366	88%	2.58×10^8
Spain	101%	90,350	18%	9.02×10^7
France	99%	301	0%	2.41×10^5
Greece	96%	1141	1%	1.01×10^6
Iran	99%	852,589	53%	8.50×10^8
Iraq	102%	371,169	84%	3.62×10^8
Israel	100%	8410	39%	1.01×10^7
Italy	100%	3571	1%	3.01×10^6
Jordan	100%	76,434	86%	9.88×10^7
Kuwait	93%	12,818	78%	1.23×10^7
Lebanon	98%	1625	16%	1.90×10^6
Morocco	92%	109,033	26%	1.19×10^8
Oman	100%	267,654	87%	2.88×10^8
Portugal	99%	9258	10%	9.15×10^6

Table A1. *Cont.*

Region or country	Calculated Country Area/FAO Area [a] [%]	Suitable Area for Solar Fuel Production [km²]	Suitable Areas/Calculated Country Area [%]	Production Volume on Suitable Areas [t/y] [b]
Qatar	97%	8147	73%	7.75×10^6
Saudi Arabia	89%	1,234,473	65%	1.35×10^9
Tunisia	99%	95,635	62%	9.75×10^7
Turkey	100%	51,497	7%	5.16×10^7
Southern Africa	100%	1,865,708	48%	2.38×10^9
South Africa	100%	638,011	52%	8.51×10^8
Namibia	100%	479,794	58%	6.61×10^8
Botswana	102%	402,197	70%	4.91×10^8
Angola	100%	345,706	28%	3.78×10^8
South America	92%	1,236,859	23%	1.47×10^9
Argentina	90	827,786	34%	9.32×10^8
Bolivia	100	210,907	19%	2.58×10^8
Chile	70	135,695	26%	2.12×10^8
Peru	101	62,471	5%	6.88×10^7
Australia	100%	4,487,022	59%	5.81×10^9
USA	84%	599,503	8%	7.77×10^8

a. Area as determined by the Food and Agriculture Organization of the United Nations (www.fao.org); b. Jet fuel; not counting the by-product naphtha.

The areas calculated add up to the areas indicated by the FAO in the case of Southern Africa and Australia but are below 100% for the other regions. In the case of Chile and the USA, this is due to the fact that these countries extend beyond the latitude boundaries of the DNI values used for the calculation and are thus out of bounds. In the case of the MED region, Saudi Arabia, the United Arab Emirates, and Yemen are underestimated with 83%–89% of the area indicated by the FAO, while the other countries achieve values above 90%. The deviations are therefore largely due to the cut-off latitude of the data, whereas in the MED region they could be due to inconsistencies between the reported FAO data and the country shapes used for the calculations. The error is however such that the available areas are slightly underestimated in single countries. When the latitude boundaries are taken into account, the calculated areas are on average very close to the reported FAO values.

Table A2. Estimates for interest rates, inflation, and the nominal weighted average cost of capital (WACC).

Country	Debt Interest Rate	Government Bond Yields	Equity Risk Premium	Inflation (5-y Average)	WACC
Algeria	0.0800	0.0575	0.1003	0.0460	0.1111
Angola	0.1791	0.0775	0.1142	0.1716	0.1841
Argentina	0.2787	0.2008	0.1142	0.0726	0.2932
Australia	0.0527	0.0262	0.0508	0.0196	0.0624
Botswana	0.0669	0.0600	0.0606	0.0390	0.0884
Chile	0.0458	0.0500	0.0578	0.0332	0.0706
Egypt	0.1884	0.1750	0.1258	0.1234	0.2333
France	0.0147	0.0082	0.0565	0.0064	0.0347
Greece	0.0550	0.0527	0.1156	−0.0046	0.1003
Iran	0.1800	0.1269	0.0842	0.1618	0.1924
Iraq	0.0400	0.0571	0.1372	0.0122	0.1017
Israel	0.0340	0.0216	0.0589	0.0022	0.0526
Italy	0.0315	0.0324	0.0727	0.0054	0.0610
Jordan	0.0840	0.0792	0.1027	0.0186	0.1231
Kuwait	0.0495	0.0267	0.0565	0.0290	0.0630
Lebanon	0.0834	0.1015	0.1258	0.0134	0.1410
Mauritania	0.1700	0.1538	0.1441	0.0244	0.2212
Morocco	0.0570	0.0383	0.0796	0.0124	0.0814
Namibia	0.0997	0.1088	0.0796	0.0542	0.1352
Oman	0.0528	0.0600	0.0727	0.0100	0.0848
Peru	0.1530	0.0575	0.0646	0.0318	0.1406

Table A2. *Cont.*

Country	Debt Interest Rate	Government Bond Yields	Equity Risk Premium	Inflation (5-y Average)	WACC
Portugal	0.0350	0.0260	0.0796	0.0058	0.0632
Qatar	0.0485	0.0482	0.0578	0.0230	0.0715
Saudi Arabia	0.0830	0.0450	0.0589	0.0162	0.0914
South Africa	0.1039	0.0950	0.0762	0.0562	0.1308
Spain	0.0220	0.0160	0.0727	0.0052	0.0487
Tunisia	0.0731	0.0575	0.1027	0.0492	0.1079
Turkey	0.1520	0.2096	0.0796	0.0860	0.2069
USA	0.0405	0.0312	0.0508	0.0132	0.0571

Table A3. Monthly labor rates in 2015 international dollars (rounded to full dollars).

Country	Managers	Engineers	Admin	Technician	Worker
Algeria	960	594	309	387	226
Angola	781	1446	379	790	388
Argentina	1759	1015	917	722	415
Australia	8250	6043	3916	4098	3005
Bahrain	3177	2312	1429	1303	846
Bolivia	799	624	401	392	275
Botswana	1242	925	392	421	350
Chile	3217	1684	649	654	437
Egypt	338	444	458	192	166
France	7018	4811	2830	2850	2346
Greece	1802	1197	974	856	681
Iran	1068	666	375	378	210
Iraq	1131	635	385	298	215
Israel	4564	3603	1864	2296	1088
Italy	9231	4219	3028	2685	2279
Jordan	2058	1069	652	671	463
Kuwait	3063	2218	1346	991	692
Morocco	1158	646	322	471	320
Mozambique	360	353	359	160	149
Namibia	916	637	372	356	304
Oman	2241	1528	923	852	669
Peru	6369	1886	1118	1931	1292
Portugal	1655	1404	813	761	548
Qatar	7782	5697	4257	1127	881
Saudi Arabia	3260	2889	1857	1284	798
South Africa	2703	2697	877	719	296
Spain	4483	3159	1958	1919	1205
State of Palestine	1385	915	532	583	465
Tunisia	969	747	457	483	332
Turkey	2980	2084	931	790	642
UAE	6001	3727	1945	651	489
USA	9056	6397	4111	4196	2747

Appendix A.1. Cost-Supply Curves per Region

Appendix A.1.1. MED Region

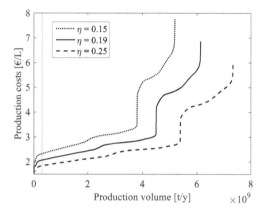

Figure A1. The cost-supply curve of solar thermochemical jet fuel for the MED region with current jet fuel demand in the selected countries [93] (**green line**) and world jet fuel demand [91] (**red line**).

Appendix A.1.2. USA

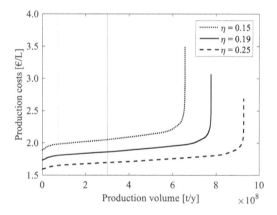

Figure A2. The cost-supply curve of solar thermochemical jet fuel for the USA with current US jet fuel demand [93] (**green line**) and world jet fuel demand [91] (**red line**).

Appendix A.1.3. Australia

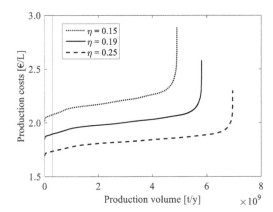

Figure A3. The cost-supply curve of solar thermochemical jet fuel for Australia with current national jet fuel demand [93] (**green line**) and world jet fuel demand [91] (**red line**).

Appendix A.1.4. South America

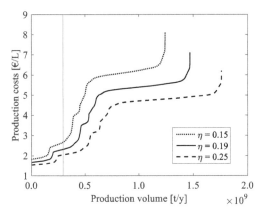

Figure A4. The cost-supply curve of solar thermochemical jet fuel for South America (Chile, Bolivia, Peru, Argentina) with current national jet fuel demand [93] (**green line**) and world jet fuel demand [91] (**red line**).

Appendix A.1.5. Southern Africa

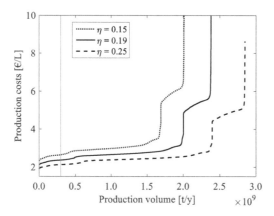

Figure A5. The cost-supply curve of solar thermochemical jet fuel for Southern Africa (South Africa, Botswana, Namibia, Angola) with current national jet fuel demand [93] (**green line**) and world jet fuel demand [91] (**red line**).

Appendix A.2. Production Costs per Region

In the following, maps of production costs for each of the regions are shown. Unsuitable areas have been removed from the maps and the production costs are indicated by color code between production costs of 1.5 €/L (dark green) and production costs of 2.5 €/L and above (dark red).

Appendix A.2.1. MED Region

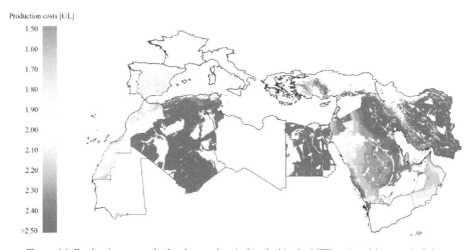

Figure A6. Production costs of solar thermochemical jet fuel in the MED region. Mauretania, Lybia, Syria, Palestine, and Yemen have been excluded from the analysis.

Appendix A.2.2. USA

Figure A7. Production costs of solar thermochemical fuel in the USA. Hawaii and Alaska have been excluded from the analysis.

Appendix A.2.3. Australia

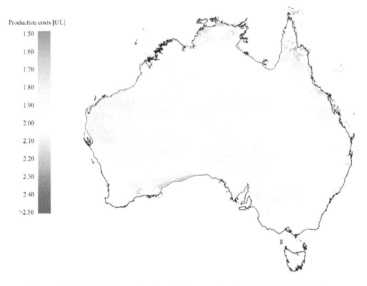

Figure A8. Production costs of solar thermochemical jet fuel in Australia.

Appendix A.2.4. South America

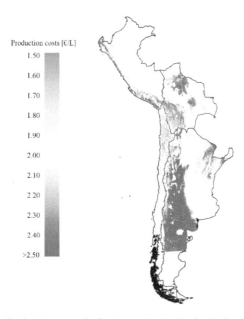

Figure A9. Production costs of solar thermochemical jet fuel in South America.

Appendix A.2.5. Southern Africa

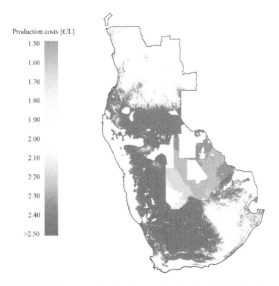

Figure A10. Production costs of solar thermochemical jet fuel in Southern Africa.

Appendix A.3. Process Efficiency

The process efficiency is defined as the lower heating value of produced jet fuel divided by the solar energy entering the system boundary, which is then converted to heat and electricity to run the process. For the production of one functional unit consisting of 1 L of jet fuel and 0.84 L of naphtha, 1.36×10^9 J of solar primary energy are required. The solar-to-heat efficiency is 0.52 [94] and the solar-to-electricity efficiency is assumed to be 0.2. The total energy input to the solar thermochemical step is 605.4 MJ per functional unit at a reactor efficiency of 19.0% (excluding vacuum pumping and gas separation). The energy requirements of the other process steps are shown in the table below. The CO_2/H_2O capture is assumed to be the Climeworks process [27] with a specific energy requirement of 300 kWh$_{el}$/t CO_2 and 1600 kWh/t CO_2. The energy requirements for the gas-to-liquids plant is modeled with an Aspen model and those for the thermochemical reactor with a Matlab model.

Table A4. Energy requirements for process steps.

Process Step	Heat Requirement [MJ/f.u.]	Electricity Requirement [MJ$_e$/f.u.]
CO_2/H_2O capture	27.7	5.2
CO_2 storage		1.1
Thermochemical reactor	605.4	
Gas separation		0.1
Syngas storage		3.6
FT conversion	0.7	1.3
Product upgrading	2.7	0.5
Steam reforming	20.6	5.1
Total	657.1	16.9

At an output of 1 L of jet fuel at a LHV of 33.4 MJ/L [95] and 0.84 L of naphtha at 31.1 MJ/L [95], the process efficiency from incident solar energy to chemical energy stored in the products is $(33.4 + 0.84 \times 31.1)/(657.1/0.52 + 16.9/0.2) = 4.4\%$ or 2.5% when counting only jet fuel.

Appendix A.4. Assumptions Regarding Process Parameters and Costs

Table A5. Reactor efficiency.

Parameter	Value	Unit
Concentration ratio	5000	-
Oxidation temperature	1000	K
Reduction temperature	1900	K
Efficiency gas heat recuperation	0.7	-
Efficiency solid heat recuperation	0.7	-
Reduction pressure	1100	Pa
Oxidation pressure	1.013×10^5	Pa
Efficiency CO_2 conversion	0.5	-

Table A6. Energy requirements, efficiencies, and costs.

Process Step	Value	Unit	Source
H_2O/CO_2 capture from atmosph			
Electricity	300	kWh/t	[96]
Heat	1500	kWh/t	[96]
Investment costs	350	€/(t y)	
O&M costs	40	€/t	
H_2O storage [a]	7.73×10^6	€	[97]
CO_2 storage (compressors) [a]	12.6×10^6	€	[98]

Table A6. *Cont.*

Process Step	Value	Unit	Source
CO_2 storage (tanks) [a]	28.9×10^6	€	[99]
Concentration of sunlight			
Optical conc. efficiency	51.6%	-	[100]
Costs of heliostats	100	€/m^2	
Costs of tower	20	€/kWth	[101]
Thermochemistry			
Costs of reactor shell	14.9	€/kWth	
Jet vacuum pumps [a,c] (inv. costs)	58.6×10^6	€	[73]
Ceria	5	€/kg	
Syngas storage [a]			
Pressure	16	Bar	
Power (H_2 compression)	14.5	MW	
Power (CO compression)	6.50	MW	
Inv. costs (H_2 compression)	9.70×10^6	€	[98]
Inv. costs (CO compression)	4.44×10^6	€	[98]
Fischer–Tropsch synthesis			
Pressure	25	bar	
Investment costs	23,000	€/bpd	[79]
O&M costs	4	€/bbl	[79]
Hydrocracking [b]			
Heat	5.2	MW	
Electricity	0.9	MW	
Steam reforming [b]			
Heat	39.9	MW	
Electricity	10.0	MW	
CO_2 capture with MEA [a]			
Heat	84.1	MW	
Electricity	0.52	MW	
Water input	1.34	L/f.u.	
Investment costs	12.9×10^6	€	[102]
Product pipeline [a]			
Investment costs	90	€/m	[81]
Ship transport			
Unit cost	8×10^{-3}	€/L	[82]
Labor costs			
Number of managers	5		
Number of engineers	30		
Number of clerks	15		
Number of technicians	100		
Number of workers	246		

Table A6. *Cont.*

Process Step	Value	Unit	Source
Salary of managers	13,890	$/y	[89]
Salary of engineers	7746	$/y	[89]
Salary of clerks	3869	$/y	[89]
Salary of technicians	5646	$/y	[89]
Salary of workers	3839	$/y	[89]

[a] The associated O&M costs are 5% of the investment costs. [b] The associated costs are assumed to be included in the cost of FT. [c] The O&M costs are included in the labor costs.

Appendix A.5. Sensitivity of Production Costs

In the following, the sensitivity of production costs with respect to a change in the inflation rate and of the weighted average costs of capital is shown. Either parameter is varied between 0% and 10%, while all others are held constant.

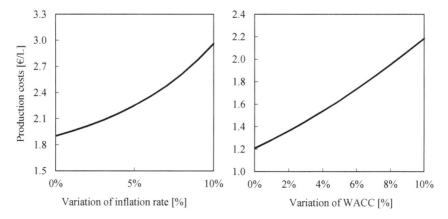

Figure A11. The sensitivity of production costs with respect to a change in the inflation rate (**left**) and weighted average costs of capital (**right**).

References

1.   U.S. Energy Information Administration (EIA). *Annual Energy Outlook 2018—with projections to 2050*; EIA: Washington, DC, USA, 2018.
2.   Unruh, G.C. Understanding carbon lock-in. *Energy Policy* **2000**, *28*, 817–830. [CrossRef]
3.   Furler, P.; Scheffe, J.; Gorbar, M.; Moes, L.; Vogt, U.; Steinfeld, A. Solar thermochemical CO_2 splitting utilizing a reticulated porous ceria redox system. *Energy Fuels* **2012**, *26*, 7051–7059. [CrossRef]
4.   Lapp, J.; Davidson, J.H.; Lipiński, W. Efficiency of two-step solar thermochemical non-stoichiometric redox cycles withheat recovery. *Energy* **2012**, *37*, 591–600. [CrossRef]
5.   Chueh, W.C.; Falter, C.; Abbott, M.; Scipio, D.; Furler, P.; Haile, S.M.; Steinfeld, A. High-Flux Solar-Driven Thermochemical Dissociation of CO2 and H2O Using Nonstoichiometric Ceria. *Science* **2010**, *330*, 1797–1801. [CrossRef] [PubMed]
6.   Marxer, D.; Furler, P.; Scheffe, J.; Geerlings, H.; Falter, C.; Batteiger, V.; Sizmann, A.; Steinfeld, A. Demonstration of the entire production chain to renewable kerosene via solar thermochemical splitting of H_2O and CO_2. *Energy Fuels* **2015**, *29*, 3241–3250. [CrossRef]
7.   Marxer, D.; Furler, P.; Takacs, M.; Steinfeld, A. Solar thermochemical splitting of CO_2 into separate streams of CO and O_2 with high selectivity, stability, conversion, and efficiency. *Energy Environ. Sci.* **2017**, *10*, 1142–1149. [CrossRef]

8. Concentrating Solar Power for the Mediterranean Region. Available online: http://www.solarec-egypt.com/resources/MED-CSP_Executive_Summary_Final.pdf (accessed on 14 January 2019).
9. Breyer, C.; Knies1, G. Global Energy Supply Potential of Concentrating. In Proceedings of the SolarPACES 2009, Berlin, Germany, 15–18 September 2009.
10. International Energy Agency. *Technology Roadmap: Concentrating Solar Power*; IEA: Paris, France, 2010.
11. European Academies Science Advisory Council. *Concentrating Solar Power: Its Potential Contribution to a Sustainable Energy Future*; EASAC: Halle, Germany, 2011.
12. Lopez, A.; Roberts, B.; Heimiller, D.; Blair, N.; Porro, G. {US} Renewable Energy Technical Potentials: A {GIS}-Based Analysis. *Contract* **2012**, *303*, 275–3000.
13. IRENA. *Estimating the Renewable Energy Potential in Africa*; IRENA: Masdar City, United Arab Emirates, 2014.
14. Trieb, F.; Schillings, C.; O'Sullivan, M.; Pregger, T.; Hoyer-Klick, C. Global potential of concentrating solar power. In Proceedings of the SolarPACES 2009, Berlin, Germany, 15–18 September 2009.
15. Journal, I.; Planning, S.E.; Vol, M. Estimation of the Global Solar Energy Potential and Photovoltaic Cost. *Int. J. Sustain. Energy Plan. Manag.* **2016**, *9*, 17–30.
16. Terwel, R.; Kerkhoven, J. Carbon neutral aviation with current technology: the take-off of synthetic fuel production in the Netherlands. Available online: https://kalavasta.com/pages/projects/aviation.html (accessed on 14 January 2019).
17. Schmidt, P.; Weindorf, W.; Roth, A.; Batteiger, V.; Riegel, F. *Power-to-Liquids Potentials and Perspectives for the Future Supply of Renewable Aviation Fuel*; Umweltbundesamt: Dessau-Roßlau, Germany, 2016.
18. König, D.H.; Freiberg, M.; Dietrich, R.-U.; Wörner, A. Techno-economic study of the storage of fluctuating renewable energy in liquid hydrocarbons. *Fuel* **2015**, *159*, 289–297. [CrossRef]
19. Schmidt, P.; Batteiger, V.; Roth, A.; Weindorf, W.; Raksha, T. Power-to-Liquids as Renewable Fuel Option for Aviation: A Review. *Chemie-Ingenieur-Technik* **2018**, *90*, 127–140. [CrossRef]
20. Huld, T.; Moner-Girona, M.; Kriston, A. Geospatial Analysis of Photovoltaic Mini-Grid System Performance. *Energies* **2017**, *10*, 218. [CrossRef]
21. Ennaceri, H.; Ghennioui, A.; Benyoussef, A.; Ennaoui, A.; Khaldoun, A. Direct normal irradiation-based approach for determining potential regions for concentrated solar power installations in Morocco. *Int. J. Ambient Energy* **2018**, *39*, 78–86. [CrossRef]
22. Long, H.; Li, X.; Wang, H.; Jia, J. Biomass resources and their bioenergy potential estimation: A review. *Renew. Sustain. Energy Rev.* **2013**, *26*, 344–352. [CrossRef]
23. Perea-Moreno, M.-A.; Samerón-Manzano, E.; Perea-Moreno, A.-J. Biomass as Renewable Energy: Worldwide Research Trends. *Sustainability* **2019**, *11*, 863. [CrossRef]
24. Deng, Y.Y.; Koper, M.; Haigh, M.; Dornburg, V. Country-level assessment of long-term global bioenergy potential. *Biomass Bioenergy* **2015**, *74*, 253–267. [CrossRef]
25. Searle, S.; Malins, C. A reassessment of global bioenergy potential in 2050. *GCB Bioenergy* **2015**, *7*, 328–336. [CrossRef]
26. Wurzbacher, J.A.; Gebald, C.; Piatkowski, N.; Steinfeld, A. Concurrent separation of CO_2 and H_2O from air by a temperature-vacuum swing adsorption/desorption cycle. *Environ. Sci. Technol.* **2012**, *46*, 9191–9198. [CrossRef]
27. Climeworks Climeworks. Available online: www.climeworks.ch (accessed on 7 January 2020).
28. Scheffe, J.R.; Steinfeld, A. Thermodynamic analysis of cerium-based oxides for solar thermochemical fuel production. *Energy Fuels* **2012**, *26*, 1928–1936. [CrossRef]
29. Bulfin, B.; Call, F.; Lange, M.; Lübben, O.; Sattler, C.; Pitz-Paal, R.; Shvets, I.V. Thermodynamics of CeO_2 Thermochemical Fuel Production. *Energy Fuels* **2015**, *29*, 1001–1009. [CrossRef]
30. Falter, C.P.; Sizmann, A.; Pitz-Paal, R. Modular reactor model for the solar thermochemical production of syngas incorporating counter-flow solid heat exchange. *Sol. Energy* **2015**, *122*. [CrossRef]
31. Falter, C.P.; Pitz-Paal, R. A generic solar-thermochemical reactor model with internal heat diffusion for counter-flow solid heat exchange. *Sol. Energy* **2017**, *144*, 569–579. [CrossRef]
32. Falter, C.P.; Pitz-Paal, R. Modeling counter-flow particle heat exchangers for two-step solar thermochemical syngas production. *Appl. Therm. Eng.* **2018**, *132*, 613–623. [CrossRef]
33. Lapp, J.; Davidson, J.H.; Lipiński, W. Heat Transfer Analysis of a Solid-Solid Heat Recuperation System for Solar-Driven Nonstoichiometric Cycles. *J. Sol. Energy Eng.* **2013**, *135*, 031004. [CrossRef]

34. Felinks, J.; Brendelberger, S.; Roeb, M.; Sattler, C.; Pitz-Paal, R. Heat recovery concept for thermochemical processes using a solid heat transfer medium. *Appl. Therm. Eng.* **2014**, *73*, 1006–1013. [CrossRef]

35. Ermanoski, I.; Siegel, N.P.; Stechel, E.B. A New Reactor Concept for Efficient Solar-Thermochemical Fuel Production. *J. Sol. Energy Eng.* **2013**, *135*, 031002. [CrossRef]

36. Welte, M.; Barhoumi, R.; Zbinden, A.; Scheffe, J.R.; Steinfeld, A. Experimental Demonstration of the Thermochemical Reduction of Ceria in a Solar Aerosol Reactor. *Ind. Eng. Chem. Res.* **2016**, *55*, 10618–10625. [CrossRef]

37. Le Gal, A.; Abanades, S. Dopant incorporation in ceria for enhanced water-splitting activity during solar thermochemical hydrogen generation. *J. Phys. Chem. C* **2012**, *116*, 13516–13523. [CrossRef]

38. Hao, Y.; Yang, C.-K.; Haile, S.M. Ceria–Zirconia Solid Solutions (Ce 1– x Zr x O 2−δ, x ≤ 0.2) for Solar Thermochemical Water Splitting: A Thermodynamic Study. *Chem. Mater.* **2014**, *26*, 6073–6082. [CrossRef]

39. Call, F.; Roeb, M.; Schmücker, M.; Bru, H.; Curulla-ferre, d.; Sattler, C.; Pitz-paal, R. Thermogravimetric Analysis of Zirconia-Doped Ceria for Thermochemical Production of Solar Fuel. *J. Anal. Chem.* **2013**, *4*, 37–45. [CrossRef]

40. Scheffe, J.R.; Weibel, D.; Steinfeld, A. Lanthanum-strontium-manganese perovskites as redox materials for solar thermochemical splitting of H_2O and CO_2. *Energy Fuels* **2013**, *27*, 4250–4257. [CrossRef]

41. Demont, A.; Abanades, S.; Beche, E. Investigation of perovskite structures as oxygen-exchange redox materials for hydrogen production from thermochemical two-step water-splitting cycles. *J. Phys. Chem. C* **2014**, *118*, 12682–12692. [CrossRef]

42. McDaniel, A.H.; Miller, E.C.; Arifin, D.; Ambrosini, A.; Coker, E.N.; O'Hayre, R.; Chueh, W.C.; Tong, J. Sr- and Mn-doped $LaAlO_3$−δ for solar thermochemical H_2 and CO production. *Energy Environ. Sci.* **2013**, *6*, 2424. [CrossRef]

43. Stechel, E.B.; Miller, J.E. Re-energizing CO_2 to fuels with the sun: Issues of efficiency, scale, and economics. *J. CO_2 Util.* **2013**, *1*, 28–36. [CrossRef]

44. Falter, C.; Batteiger, V.; Sizmann, A. Climate Impact and Economic Feasibility of Solar Thermochemical Jet Fuel Production. *Environ. Sci. Technol.* **2016**, *50*, 470–477. [CrossRef] [PubMed]

45. Steinfeld, A. Solar thermochemical production of hydrogen—A review. *Sol. Energy* **2005**, *78*, 603–615. [CrossRef]

46. QGIS Development Team (2017). QGIS Geographic Information System. *Open Source Geospatial Foundation Project.* Available online: http://qgis.osgeo.org (accessed on 7 February 2019).

47. Broxton, P.D.; Zeng, X.; Sulla-Menashe, D.; Troch, P.A. A Global Land Cover Climatology Using MODIS Data. *J. Appl. Meteorol. Climatol.* **2014**, *53*, 1593–1605. [CrossRef]

48. IUCN; UNEP-WCMC. *The World Database on Protected Areas (WDPA)*; IUCN: Gland, Switzerland; UNEP-WCMC: Cambridge, UK, 2017; Available online: www.protectedplanet.net (accessed on 10 November 2018).

49. Food and Agriculture Organiziation of the United Nations (FAO). Land and Water Development Division. *The Digital Soil Map of the World.* Available online: http://ref.data.fao.org/map?entryId=446ed430-8383-11db-b9b2-000d939bc5d8&tab=metadata (accessed on 10 November 2018).

50. Global Administrative Areas Database. Available online: https://gadm.org/ (accessed on 14 January 2019).

51. Wessel, P. Global Self-Consistent, Hierarchical, High-resolution Geography Database, v2.3.7. Available online: http://www.soest.hawaii.edu/pwessel/gshhg/ (accessed on 15 November 2019).

52. National Oceanic and Atmospheric Administration (NOAA). Digital Elevation Model. Available online: https://www.ngdc.noaa.gov/mgg/global/global.html (accessed on 14 January 2019).

53. World Bank. Global Solar Atlas. Available online: www.globalsolaratlas.info (accessed on 8 May 2018).

54. Short, W.; Packey, D.J.; Holt, T. *A Manual for the Economic Evaluation of Energy Efficiency and Renewable Energy Technologies*; NREL: Golden, CO, USA, 1995.

55. World Economic Forum. *The Global Competitiveness Report 2017-2018*; World Economic Forum: Geneva, Switzerland, 2018; ISBN 978-1-944835-11-8.

56. Duarte, F.; Rosa, C. The equity risk premium: A review of models. In *Federal Reserve Bank of New York Staff Reports*; New York Fed: New York, NJ, USA, 2015.

57. Trading Economics. Government Bond yields. Available online: https://tradingeconomics.com/bonds (accessed on 2 August 2019).

58. World Government Bonds. 10 Year Bond Yields. Available online: http://www.worldgovernmentbonds.com/ (accessed on 2 August 2019).
59. Damodoran, A. Estimated Country Risk Premiums. Available online: http://pages.stern.nyu.edu/ ~{}adamodar/ (accessed on 3 July 2019).
60. Damodoran, A. Equity Risk Premiums (ERP): Determinants, Estimation and Implications. Available online: http://people.stern.nyu.edu/adamodar/pdfiles/papers/ERP2011.pdf (accessed on 3 July 2019).
61. International Monetary Fund. Lending Rates, Percent per Annum. Available online: https://data.imf.org/ regular.aspx?key=61545867 (accessed on 3 July 2019).
62. CEIC. Bank Lending Rate. Available online: https://www.ceicdata.com/en/indicator/bank-lending-rate (accessed on 3 July 2019).
63. Central Intelligence Agency. Country Comparison: Commercial Bank Prime Lending Rate. Available online: https://www.cia.gov/library/publications/the-world-factbook/fields/231rank.html (accessed on 4 July 2019).
64. International Monetary Fund. Inflation Rate, Average Consumer Prices. Available online: https://www.imf. org/external/datamapper/PCPIPCH@WEO/OEMDC (accessed on 5 October 2019).
65. White, I.R.; Royston, P.; Wood, A.M. Multiple imputation using chained equations: Issues and guidance for practice. *Stat. Med.* **2011**, *30*, 377–399. [CrossRef]
66. World Bank. GDP per Capita (Current US$). Available online: https://data.worldbank.org/indicator/NY.GDP. PCAP.CD (accessed on 5 October 2019).
67. Labordena, M.; Patt, A.; Bazilian, M.; Howells, M.; Lilliestam, J. Impact of political and economical barriers for concentrating solar power in Sub-Saharan Africa. *Energy Policy* **2017**, *102*, 52–72. [CrossRef]
68. SunShot U.S. Department of Energy SunShot Vision Study. *U.S. Dep. Energy* **2012**, 69–96.
69. Mehos, M.; Turchi, C.; Jorgenson, J.; Denholm, P.; Ho, C.; Armijo, K. On the path of SunShot: Advancing Concentrating Solar Power Technology, Performance, and Dispatchability. *SunShot* **2016**, 1–66. [CrossRef]
70. IRENA. *Renewable Power Generation Costs in 2017*; IRENA: Masdar City, United Arab Emirates, 2018.
71. International Air Transport Association. Jet Fuel Price Development. Available online: http://www.iata.org/ publications/economics/fuel-monitor/Pages/price-development.aspx (accessed on 30 March 2016).
72. Historic Naphtha Prices. Available online: http://www.finanzen.net/rohstoffe/naphtha/historisch (accessed on 17 February 2015).
73. Brendelberger, S.; von Storch, H.; Bulfin, B.; Sattler, C. Vacuum pumping options for application in solar thermochemical redox cycles – Assessment of mechanical-, jet- and thermochemical pumping systems. *Sol. Energy* **2017**, *141*, 91–102. [CrossRef]
74. Climeworks. Capturing CO_2 from Air. In *Manuf. Green Fuels from Renew. Energy Work. DTU Risø, Roskilde, 14 April 2015*; Climeworks: Zürich, Switzerland, 2015.
75. Chueh, W.C.; Haile, S.M. A thermochemical study of ceria: exploiting an old material for new modes of energy conversion and CO_2 mitigation. *Philos Trans. A Math Phys Eng Sci* **2010**, *368*, 3269–3294. [CrossRef] [PubMed]
76. Lilliestam, J.; Ollier, L.; Pfenninger, S. The dragon awakens: Will China save or conquer concentrating solar power? Available online: https://aip.scitation.org/doi/pdf/10.1063/1.5117648?class=pdf (accessed on 11 October 2019).
77. pv magazine. Los Angeles Seeks Record Setting Solar Power Price under 2¢/kWh. Available online: https://pv-magazine-usa.com/2019/06/28/los-angeles-seeks-record-setting-solar-power-price-under-2¢-kwh/ (accessed on 11 October 2019).
78. pv magazine. Mexico's power auction pre-selects 16 bids with average price of $20.57/MWh and 2.56 GW of combined capacity. Available online: https://www.pv-magazine.com/2017/11/16/mexicos-power-auction-pre-selects-16-bids-with-average-price-of-20-57mwh-and-2-56-gw-of-combined-capacity/ (accessed on 11 October 2019).
79. Velocys. *Private Communication 2013*; Velocys: Oxford, UK, 2013.
80. Smith, C.E. Crude oil pipeline growth, revenues surge; construction costs mount. *Oil Gas J.* **2014**. Available online: https://www.ogj.com/pipelines-transportation/article/17210347/crude-oil-pipeline-growth-revenues-surge-construction-costs-mount (accessed on 11 October 2019).
81. Peters, M.S.; Timmerhaus, K.D.; West, R.E. *Plant Design and Economics for Chemical Engineers*; McGraw-Hill Education: New York, NJ, USA, 2001.

82. US Department of Transportation. Average Freight Revenue. Available online: https://www.bts.gov/archive/publications/national_transportation_statistics/2000/3-17 (accessed on 18 January 2019).

83. Wallasch, A.-K.; Lüers, S.; Vidican, G.; Breitschopf, B.; Richter, A.; Kuntze, J.-C.; Noll, J. The Socio-economic Benefits of Solar and Wind Energy. Available online: https://www.irena.org/-/media/Files/IRENA/Agency/Publication/2014/Socioeconomic_benefits_solar_wind.pdf (accessed on 18 January 2019).

84. Sooriyaarachchi, T.M.; Tsai, I.T.; El Khatib, S.; Farid, A.M.; Mezher, T. Job creation potentials and skill requirements in, PV, CSP, wind, water-to-energy and energy efficiency value chains. *Renew. Sustain. Energy Rev.* **2015**, *52*, 653–668. [CrossRef]

85. IRENA. *Renewable Energy Jobs: Status, Prospects & Policies*; IRENA: Masdar City, United Arab Emirates, 2012.

86. Applied Analysis. *Large-Scale Solar Industry. Economic and Fiscal Impact Analysis*; Applied Analysis: Las Vegas, NV, USA, 2009.

87. World Bank. *Middle East and North Africa Region Assessment of the Local Manufacturing Potential for Concentrated Solar Power (CSP) Projects*; World Bank: Washington, DC, USA, 2011; Volume 1, Chapter 4.

88. SolarPACES. SolarReserve Breaks CSP Price Record with 6 Cent Contract. Available online: http://www.solarpaces.org/solarreserve-breaks-csp-price-record-6-cent-contract/ (accessed on 22 October 2018).

89. ILO. Mean nominal monthly earnings of employees by sex and occupation—Harmonized series. Available online: https://www.ilo.org/shinyapps/bulkexplorer51/?lang=en&segment=indicator&id=EAR_4MTH_SEX_ECO_CUR_NB_A (accessed on 23 October 2018).

90. Kim, J.; Miller, J.E.; Maravelias, C.T.; Stechel, E.B. Comparative analysis of environmental impact of S2P (Sunshine to Petrol) system for transportation fuel production. *Appl. Energy* **2013**, *111*, 1089–1098. [CrossRef]

91. IATA. *Economic Performance of the Airline Industry 2018*; Intergovernmental Panel on Climate Change, Ed.; Cambridge University Press: Cambridge, UK, 2018.

92. Food and Agriculture Organizition of the United Nations (FAO). Country profiles of the FAO. Available online: http://www.fao.org/countryprofiles/en/ (accessed on 17 January 2019).

93. U.S. Energy Information Administration. International Energy Statistics. Available online: www.eia.gov/beta/international/data/browser (accessed on 24 January 2019).

94. Pitz-Paal, R.; Botero, N.B.; Steinfeld, A. Heliostat field layout optimization for high-temperature solar thermochemical processing. *Sol. Energy* **2011**, *85*, 334–343. [CrossRef]

95. Stratton, R.W.; Wong, H.M.; Hileman, J.I. *Life Cycle Gas Emissions from Alternative Jet Fuels*; PARTNER Project 28 report; MIT: Cambridge, MA, USA, 2010; Available online: http://web.mit.edu/aeroastro/partner/reports/proj28/partner-proj28-2010-001.pdf (accessed on 24 January 2019).

96. Climeworks LLC. CO$_2$ Air Capture Demonstration Plant. Available online: http://www.climeworks.com/co2-capture-plants.html (accessed on 1 June 2014).

97. Wernet, G.; Bauer, C.; Steubing, B.; Reinhard, J.; Moreno-Ruiz, E.; Weidema, B. The ecoinvent database version 3 (part I): Overview and methodology. *Int. J. Life Cycle Assess.* **2016**, *2*, 1218–1230. [CrossRef]

98. McCoy, S.T. The Economics of CO$_2$ Transport by Pipeline and Storage in Saline Aquifers and Oil Reservoirs. Ph.D. Thesis, Carnegie Mellon University, Pittsburgh, PA, USA, 2009.

99. James, B.D.; Houchins, C.; Huya-Kouadio, J.M.; Desantis, D.A. *Final Report: Hydrogen Storage System Cost Analysis*; OSTI.gov: Oak Ridge, TN, USA, 2016. Available online: https://www.osti.gov/servlets/purl/1343975 (accessed on 24 January 2019).

100. Sargent & Lundy. *Assessment of Parabolic Trough and Power Tower Solar Technology*; SL-5641; Sargent & Lundy: Chicago, IL, USA, 2003.

101. Mancini, T.R.; Gary, J.A.; Kolb, G.J.; Ho, C.K. *Power Tower Technology Roadmap and Cost Reduction Plan*; Sandia National Laboratories: Albuquerque, NM, USA; Livermore, CA, USA, 2011.

102. Kim, J.; Johnson, T.A.; Miller, J.E.; Stechel, E.B.; Maravelias, C.T. Fuel production from CO$_2$ using solar-thermal energy: System level analysis. *Energy Environ. Sci.* **2012**, *5*, 8417. [CrossRef]

MDPI

St. Alban-Anlage 66

4052 Basel

Switzerland

Tel. +41 61 683 77 34

Fax +41 61 302 89 18

www.mdpi.com

Energies Editorial Office

E-mail: energies@mdpi.com

www.mdpi.com/journal/energies

Lightning Source UK Ltd.
Milton Keynes UK
UKHW021854221220
375650UK00003B/173